Calculator Calculus

George M⊆Carty

Calculator Calculus

George McCarty
University of California, Irvine

EduCALC Publications
Box 974, Laguna Beach, CA 92652

CALCULATOR CALCULUS

LIBRARY OF CONGRESS CATALOG CARD No. 75-27363

ISBN 0-936356-00-6 (previously 0-471-27596-4)

Revised Softbound Edition: 6th Printing, February, 1982

CONTENTS

PREFACE

How This Book Differs

This book is about the calculus. What distinguishes it, however, from other books is that it uses the pocket calculator to illustrate the theory. A computation that requires hours of labor when done by hand with tables is quite inappropriate as an example or exercise in a beginning calculus course. But that same computation can become a delicate illustration of the theory when the student does it in seconds on his calculator. † Furthermore, the student's own personal involvement and easy accomplishment give him reassurance and encouragement.

The machine is like a microscope, and its magnification is a hundred millionfold. We shall be interested in limits, and no stage of numerical approximation proves anything about the limit. However, the derivative of $f(x) = 67.89^x$, for instance, acquires real meaning when a student first appreciates its values as numbers, as limits of

---

† A quick example is 1.1^{10}, 1.01^{100}, 1.001^{1000}, ...
Another example is $t = 0.1, 0.01, ...$ in the function $(\sqrt{3t+9}-3)/t$.

difference quotients of numbers, rather than as values of a function
that is itself the result of abstract manipulation.

Similarly, the fun and excitement a student has in calculating for
himself some approximations to a few definite integrals, such as
$\int_0^1 \sqrt{1-x^2}\ dx$, will give reality to their definition as limits of
Riemann sums. When our usual algebraic manipulation of the sums for
the integrands 1, x, x^2, and perhaps x^3 is augmented by such calculations, the Fundamental Theorem of the Calculus is seen in a new light.
Instead of being misunderstood to be part of the definition of the
integral, it becomes a genuine theorem that usefully relates two dis-
parate mathematical objects.

This is not a manual of machine usage, but the student who works
through this book will gain calculational competence and a skill at
coaxing the most from his machine. Although it is not a workbook of
numerical analysis, this book will introduce that subject--there are
discussions and examples of errors, numerical quadrature, finding
zeros, evaluating functions, and solving differential equations
numerically.

The student learns respect here for calculation in problems where
theoretical methods fail and only numerical solutions exist. However,
in other problems, after he labors to form a few partial sums for a
series like 1 - 1/3 + 1/5 - ... , he will appreciate the ease and
power with which the theory gives the limiting value. Perhaps now
the calculator's buttons and twinkling lights can seduce the student
to a balanced understanding of the theory and practice of the
calculus.

There has been no attempt to be complete in the exposition of theory
in this book, but the most important theorems are cited explicitly
and illustrated numerically. The student may use these citations as
signals for review in his conventional calculus text. The chapters
are short; each one stands as independently as the underlying theory
will allow. Discussions and detailed solutions for several *Examples*
are included in each chapter. In addition there are both *Exercises*
and *Problems*. The Exercises are easy and to the point. Some Exercises include applications drawn from the biological, social, and
physical sciences. The Problems are more difficult or longer, often

they explore less central topics, and some ask for proofs. Answers to starred Exercises and Problems are given at the end of each chapter.

Classroom Use in a First Calculus Course

When this book is used as a workbook or problem manual for an introductory course that also has a conventional text, the instructor may concentrate on the demonstrative Examples and Exercises. Much of the explanatory material in this book may be left for the student to read as he studies. Roughly half of all assigned exercises might be chosen from this book and half from the conventional text. The author usually devotes one of his three weekly lectures plus one of the two weekly problem sessions to material involving the calculator.

The first two chapters in this book include topics outside the usual preliminary material for the calculus. The algorithms for square roots and for successive substitutions serve to introduce functions and graphs. They also accustom students to their machines and start them thinking about limiting processes. But Chapter 1 may be omitted entirely, and Chapter 2 may be omitted if the method of successive substitutions is briefly explained when it is needed in Exercises in Chapters 6 and 8.

Classroom Use for Other Courses

This book can serve as the text for an advanced undergraduate course. For such a course it will be appropriate to consider many of the more difficult Problems. It can also be used for a one-semester or one-quarter course having first-year calculus as prerequisite or corequisite. At this lower level, most students will find enough challenge in the Exercises, with few Problems attempted.

The author feels it is important to plan the schedule of such a course so that the material on series, in Chapters 11, 12, and 13, is sure to be covered. If necessary, Chapters 9 or 10 or both, on volumes and on curves and polar coordinates, may be omitted with no loss of continuity. Chapter 14 may also be omitted in a lower-level course, particularly if students have had no previous preparation in differential equations.

Acknowledgments

I have received advice and encouragement from several colleagues during the writing of this book. I am grateful for such help to Donald Albers, John Grover, J. L. Kelley, and Howard Tucker. The book has also benefited from the comments and criticisms of Brent Gloege and other students too numerous to name at the University of California, Irvine. Finally, I would like to thank Karen Thomas and Mary Green for their time and careful attention to detail in the preparation of the various stages of this book.

NOTE TO THE STUDENT

WHICH MACHINE?

There is an enormous variety of pocket calculators available with
features and functions in many different combinations. Many of
these calculators are suitable for learning the calculus with this
book. *The recommended machine for our work is one which has buttons
to calculate trig and log functions, displays at least eight digits,
and has an adapter-recharger.* Of course, methods for approximating
logarithmic and trigonometric functions are given in this book.
Nevertheless, it is our experience that a student who attempts to
do the work with a four- or seven-function machine will become
distracted by the copious arithmetic and eventually will de-
spair.

Most calculators that satisfy our minimal requirements also have
square root and reciprocal functions, a memory, and the internal
constant π. Most will also accept arguments for trig functions in
radians as well as degrees. In addition, some such calculators
have the ability to convert a result into "scientific notation,"

with a mantissa and an exponent. These machines also offer superior logical systems called either "Parenthetical" or "Reverse-Polish." These systems are very useful; they enable the machine to accept more complex formulas without the user having to rewrite equations or write down intermediate results. If you can afford it, we recommend that you use a machine having scientific notation and such a system of logic.

There are even more elaborate calculators. Some have multiple memories, and some may be programmed to perform repetitive computations automatically. Such features could be useful in our work, but they will not be necessary.

How to Get Started

Before beginning your study of this book, you must become familiar with the machine you will be using. You can do this by reading the explanations in the owner's manual and working out the simple examples given there. This may take a few hours if it is your first calculator. You should also test your understanding by trying to do simple arithmetic on the machine, using numbers like 2, 3, and 4 and working out problems for which you already know the answer.

Some suggestions on "Invisible Registers" and simple arithmetic, as well as the evaluation of more complicated expressions, are offered in the Appendix of this book.

What to Do When You Are Baffled

You want to learn to use the calculus. You cannot do that by osmosis, by watching someone else do it, any more than you could learn to play chess, football, or the violin by close observation. Do the Exercises. They are not repetitive drills. You can expect some of the joy of discovery and creation with each solution you construct.

When you have trouble understanding a topic, reexamine the worked-out Examples and then do some Exercises. Do the easiest ones first. With experience you will come to know which you can do in your head and which are difficult for you. Calculate and write out solutions for the harder ones. You will find that you, too, can learn the calculus and enjoy doing it.

1

SQUARES, SQUARE ROOTS, AND THE QUADRATIC FORMULA

INTRODUCTION

Since people began using numbers for measuring lengths, they have
wanted to find the square roots of those numbers. There are many
situations in which a square root is needed. For instance, knowing
the square root is useful if you want to find the length of the side
of a square field of a given area or the length of its diagonal when
the side is known. The ancient Babylonian mathematicians were even
solving quadratic equations in the time of Hammurabi, a Babylonian
king of the eighteenth century B.C. Their method for approximating
square roots was a first step in the more accurate repetitive process
discovered by Hero of Alexandria about the time of Christ. This process, now known as Newton's method, is universally used today.

The repetitive method that you will learn about in this chapter
is brand new, although it too is based on the work of the Babylonians.
Our process differs, however, since it has been developed especially
for pocket calculators.

1

Our first experience with this new sort of calculation will be exciting. It will be an easy example of what is called an algorithm. This beginning study will provide practice in the elementary operations on a calculator. This first chapter will also include some review and some new understanding of graphs and functions. Exercises will display the properties and limitations of our algorithm and apply it in the quadratic formula. In the Problems you can explore this algorithm and also Newton's method more deeply and compare their speeds of convergence.

THE DEFINITION

It is easy to square a given number x. And every squared number $x^2 = x \times x$ is positive except for 0^2. The problem of reversing this squaring operation is the problem of finding square roots, and it is not quite so simple. That is, the *square root* $\sqrt{x}$ of a given positive number x is by definition the unique positive number y for which $y^2 = x$. If we find $\sqrt{x}$, then $-\sqrt{x}$ is the other number whose square is x.

Though your calculator finds $\sqrt{x}$ at the touch of a button, *ignore that button for this chapter* and learn about the iterative methods by which the machine itself does such computations.

EXAMPLE: $\sqrt{67.89}$

Suppose we want to compute $\sqrt{67.89}$. We can first make a rough guess at the answer of 8, since $8^2 = 64$, which is fairly close to 67.89. Now we use an arithmetic trick to improve our guess. We let y stand for the number we want to find, so that $y^2 = 67.89$. Then we write

$$y^2 = 67.89$$
$$y^2 + 8y = 67.89 - 64 + 8y + 64$$
$$y(y+8) = 3.89 + 8(y+8)$$
$$y = \frac{3.89}{y+8} + 8.$$

This equation will be satisfied by y and by no other number. Since we wish to improve our guess for $\sqrt{67.89}$, we will experiment by regarding the *two appearances* of the number y in this equation as *two different numbers* that are related by the equation. That is, we relabel our guess g as $g = y_0$ and calculate

$$y_1 = \frac{3.89}{y_0 + 8} + 8.$$

To do this on your own machine, first key in $y_0 = 8.$, next add 8 to get 16., then find the reciprocal $1/16 = 0.0625$. Multiply this reciprocal by 3.89 to see 0.243125, and finally add 8 to obtain $y_1 = 8.243125$. Now the square of our guess $g = y_0$ was 64; the square of this number y_1 is 67.949110. Hence y_1 is not the square root of 67.89, but it certainly is a better approximation of it than is $y_0 = 8$.

Encouraged by this fact, we again use our equation. This time we calculate a new estimate

$$y_2 = \frac{3.89}{y_1 + 8} + 8.$$

This procedure is easy. Since y_1 is still showing in the machine, we merely add 8, reciprocate, multiply by 3.89, and again add 8. The displayed result is $y_2 = 8.2394859$. This is an even better estimate, since $y_2{}^2 = 67.889128$.

We now repeat this process methodically, calculating y_3 from y_2 and, more generally, calculating y_{i+1} from y_i for successive integers $i = 2, 3, 4,$ and 5 by the rule

$$y_{i+1} = \frac{3.89}{y_i + 8} + 8.$$

The results are shown in Table 1.1. We suggest that you duplicate our calculations and check your work against this table.

TABLE 1.1

$$y_0 = 8.$$
$$y_1 = 8.2431250$$
$$y_2 = 8.2394859$$
$$y_3 = 8.2395396$$
$$y_4 = 8.2395388$$
$$y_5 = 8.2395388$$

Notice that $y_4 = y_5$ exactly. Doesn't this mean that $y_5 = y_6$, $y_6 = y_7$, and so forth? Thus our method cannot improve the guess any further. But it has done its job already. We calculate that $y_4^2 = 67.89$ exactly.

Our machine rounds off to the nearest 8-digit answer; your calculator may display 67.889999 here. In fact, there is no 8-digit number whose square is 67.89, as you may verify by experiment. That is, the next number in the machine that is larger than y_4 is 8.2395389 and its square is 67.890001. This means that, to the limit of our machine's 8-digit accuracy, y_4 is the correct square root of 67.89. In this book, we use the equality symbol to express this, so we write $\sqrt{67.89} = 8.2395388$.

Notice that the final result $\sqrt{67.89}$ was calculated without having to jot down any intermediate result for later reentry into the machine, even on a machine without a memory button. We do, of course, repeatedly reenter the numbers 8 and 3.89, having written them down at the start. If your machine has a memory, you may find it useful to store 3.89 there and recall it as needed.

To display the process of convergence, the process of the numbers y_i becoming ever more accurate approximations to $\sqrt{67.89}$, we tabulate y_i^2 in Table 1.2. This table shows the progressive increase in the accuracy of the estimates. We emphasize, however, that the numbers y_i^2 need not be calculated. Instead, we may proceed through the iterative process of guessing $g = y_0$ and calculating in succession y_1, y_2, y_3, y_4, and y_5. We observe that $y_5 = y_4$ so that y_4 is

4

TABLE 1.2

	y_i	y_i^2
y_0	8.	64.
y_1	8.2431250	67.949109
y_2	8.2394859	67.8889128
y_3	8.2395396	67.8900012
y_4	8.2395388	67.89
y_5	8.2395388	67.89

the answer y that we sought. It is wise, as a check, to square only
the result y_4.

THE ALGORITHM

The method we have used to find $\sqrt{67.89}$ may be generalized to get a
technique called an *algorithm*, an iterative procedure for finding $\sqrt{x}$.
As before, we seek a positive number y such that $y^2 = x$. We make a
guess g for y and write

$$y^2 = x$$
$$y^2 + gy = x - g^2 + gy + g^2$$
$$y(y+g) = x - g^2 + g(y+g)$$
$$y = \frac{x-g^2}{y+g} + g.$$

As in the example, we regard the two appearances of y in the last
equation as two different numbers y_{i+1} and y_i, to get an algorithm[†]:

$$y_{i+1} = \frac{x-g^2}{y_i+g} + g.$$

[†]This method is a new one as far as we know.

This is the recipe for calculating $\sqrt{x}$: we let $y_0 = g$ and calculate $y_1 = (x-g^2)/(y_0+g) + g$. Continuing, we calculate y_2 from y_1, y_3 from y_2, and so on, until we get to a stage where $y_{n+1} = y_n$. Then we stop. The answer is $\sqrt{x} = y_n$.

EXAMPLE: $\sqrt{100}$

In order to exhibit our algorithm in action, we make the foolish guess of $g = 11$ for $\sqrt{100}$. The algorithm in this case is

$$y_{i+1} = \frac{-21}{y_i+11} + 11.$$

Results are listed in Table 1.3. This time we have omitted some

TABLE 1.3

$$y_0 = 11.$$
$$y_1 = 10.045455$$
$$y_2 = 10.002160$$
$$y_3 =$$
$$y_4 =$$
$$y_5 = 10.$$
$$y_6 = 10.$$

intermediate results. You should work out these steps for yourself and fill in the missing numbers to complete the table.

EXERCISES

1. In each case, find the square root of x by means of the algorithm of this chapter, starting with the given guess g. For which n is y_n first equal to y_{n+1}?

*a. $x = 26$ and $g = 5$ e. $x = 50.12$ and $g = 7$

*b. $x = 35$ and $g = 5$ f. $x = 63$ and $g = 7$

*c. $x = 1.11$ and $g = 1$ g. $x = 4.567$ and $g = 2$

*d. $x = 150$ and $g = 12$ h. $x = 650$ and $g = 25$

---

*Answers to starred examples and problems can be found at the end of this chapter.

2. Calculate $\sqrt{67.89}$ using a poor guess, say g = 6, to start, so y_0 = 6 and $x - g^2$ = 31.89. How many iterations are required in this case to arrive at the correct result y_1 = 8.2395388?

We tested one 8-digit machine that ended this iteration by cycling back and forth between 8.2395358 and 8.2395390. This cycling was due to loss of information after finding the reciprocal $\frac{1}{8.2395390 + 6}$. This number is approximately 0.0702269926, but it was recorded as 0.0702269 by an 8-digit machine, of course. It is fairly easy to achieve greater accuracy on such a machine: simply *divide* by 31.89 *before* reciprocating instead of *multiplying* by 31.89 *after* reciprocating. The numerical principle involved is that the least information is lost in reciprocation for numbers nearest to 1.

*3. Calculate $\sqrt{35}$ by guessing g = 6, so y_0 = 6 and $x - g^2$ = -1. Notice that $x - g^2$ is negative this time. This is okay; you must simply remember to "change its sign" to negative every time you key it into your machine. What is $\sqrt{35}$? How many iterations did it take to find it?

Suppose that to find $\sqrt{36}$ you guess g = 6, so that the guess is exact this time. What happens in the algorithm for this case?

*4. The *quadratic formula*

$$r = \frac{-b \pm \sqrt{b^2 - 4ac}}{2a}$$

gives the roots, if any, of the quadratic equation $ax^2 + bx + c = 0$. If the *discriminant* $\Delta = b^2 - 4ac$ is zero, then there is one "double" root; If $\Delta < 0$, then there are no real roots; if $\Delta > 0$, then there are two roots corresponding to the choice of signs ± before the radical. Find the roots of $x^2 - x - 1 = 0$. Then check your answers r_1 and r_2 by calculating $r_1{}^2 - r_1 - 1$ and $r_2{}^2 - r_2 - 1$.

5. Use the results of Exercise 4 to graph the function $f(x) = x^2-x-1$.
Consider values of x from -2 to +3 and use increments of 0.2, so you
will compute $f(-2.0)$, $f(-1.8)$, $f(-1.6)$, ..., up to $f(3)$. Calculate
the values $f(x)$ to 2 decimal places and interpolate to graph them on
standard graph paper. Use a whole sheet and place your origin a
little below the center of the page. Label all calculated points,
the roots, and the y-intercept $f(0)$.

6. Using the methods of Exercise 4, find the roots of $3x^2+2x-7 = 0$.
Check your answers by evaluation.

7. Graph the function $f(x) = 3x^2+2x-7$ by the methods of Exercise 5.
Use the roots found in Exercise 6.

*8. Use the quadratic formula (Exercise 4) to solve
$\frac{\pi x^2}{4} - \frac{7x}{47} - 1.729 = 0$ ($\pi = 3.1415927$). Check your answers. Sketch
a graph displaying this function of x with its x- and y- intercepts
(those are the places where the graph crosses the x- and y- axes).

9. Make a graph of the square root function $f(x) = \sqrt{x}$ as follows:
the points (x,y) that we wish to plot are those for which $y = \sqrt{x}$.
But if both x and y are positive ($0 = \sqrt{0}$, remember), then $y = \sqrt{x}$ if
and only if $x = y^2$. To plot some points, then, let $y = 0.0, 0.1, 0.2$,
0.3, ..., and so on, up to $y = 2.0$ in increments of 0.1. Square
these values of y to find the x coordinate for which $\sqrt{x} = y$. Plot
these points and sketch a curved line through them. Then use your
graph to estimate to 2 decimal places $\sqrt{.35}$, $\sqrt{0.9}$, $\sqrt{1.2}$, $\sqrt{2}$.

10. The circle of radius 4 that is centered at (2,3) has as its
equation

$$(x-2)^2 + (y-3)^2 = 4^2.$$

Find the coordinates of the points of intersection of this circle
with the coordinate axes and also the points of intersection of the
circle with the straight line $y = 4x+3$.

11. When the number x is between 0 and 1, it may be confusing to guess at its square root, that is, to find a good g to begin the calculation. One trick is to multiply x by 100, make a guess G for $\sqrt{100x}$, then use $G/10 = g$ as a guess for $\sqrt{x}$. Do this to make guesses for $\sqrt{0.5}$ and $\sqrt{0.05}$ and then find these two square roots. What could you do to guess $\sqrt{0.005}$ and $\sqrt{0.0005}$?

*12. Solve the ancient problem of "squaring the circle". In other words, find $\sqrt{\pi}$. Here π means 3.1415927, and the number $y = \sqrt{\pi}$ which you find is the length y of the side of a square whose area y^2 is π, the area of a circle of radius 1.

   (Of course, both π and $\sqrt{\pi}$ are irrational numbers, though a calculator can only deal with decimal fractions that approximate these numbers.)

*13. Suppose two animals are similarly shaped, and r is the ratio of their linear dimensions. For example, if one is twice as high as the other and twice as long, then $r = 2$. Their surface areas will then bear the ratio r^2 and their volumes the ratio r^3. Since weights are proportional to volumes, their weights also have the ratio r^3. Suppose that a lab has been  using 110-gram mice in a study. If the experiment now calls for mice with a surface area 25% greater than before, what weight mice are needed? (Hint: the ratio of surface areas is $r^2 = 1.25$. Use the algorithm. Round off your answer to the nearest gram.)

*14. Suppose a savings institution offers a deposit certificate that costs $1000 and pays back $1200 at the end of its two-year term. This is 20% simple interest over the two years. But the local bank usually pays yearly interest, which it adds to the accounts at the end of each year. This is called *annual compounding* of the interest. What rate $r\%$ of annually compounded interest would yield the same 20% return after two years? (Hint: $(1 + r/100)^2 = 1200/1000$. Use the algorithm. Round off your answer to the nearest 1/4 percentage point.)

PROBLEMS

P1. Show graphically how the formula $y_{i+1} = \dfrac{3.89}{y_i + 8} + 8$ works. Begin by making a graph showing that both the functions $Y = X$ and $Y = 3.89/(X+8) + 8$ in a very large scale, using values of X from 0 to 16. Do you see why the intersection of the two graphs gives $\sqrt{67.89}$? Now, make another graph of these two functions that magnifies the region near their intersections. Use values of X from 7.9 to 8.3 in increments of 0.005. On this graph plot the points $(8,0)$, $(8,y_1)$, (y_1,y_1), (y_1,y_2), (y_2,y_2) and connect these points with a dotted line from each one to the subsequent one. Do you now understand how this dotted line depicts the convergence of our process to its limiting value?

P2. "Prove" that it is not necessary to check your result when you have obtained a square root by the method of our example. That is, give a reasoned argument using arithmetic to show that when you calculate y_{n+1} and find it is equal to y_n, then y_n must be the correct answer, that is, $y_n{}^2$ must be equal to x. (However, it is prudent, reassuring, and fun to check answers. You may have made an error in calculating the number $x - g^2$, for instance, that you used repeatedly to obtain y_n. Your answer would then be wrong even though $y_n = y_{n+1}$.)

P3. Study the operation of our algorithm more carefully in two ways, using $\sqrt{67.89}$ as an example.

First, suppose that in your calculation of y_2 from y_1 you mistakenly key in 3.59 instead of 3.89. Pretend not to notice your error, which will of course result in an incorrect value for y_2. Then continue without making any further mistakes to follow the algorithm, calculating y_3 from your erroneous value y_2, y_4 from y_3, and so on, tabulating these results as you go. Finally, compare your table with Table 1.1. Do you see that our algorithm is "self-correcting"?

Second, follow the algorithm to derive y_1 from $y_0 = g = 8$. Then regard y_1 as a new and better guess g_1 and calculate

$$g_2 = (67.89 - g_1{}^2)/2g_1 + g_1.$$

Derive g_3 from g_2 in a similar fashion. Continue this new iterative algorithm until a stage where $g_{i+1} = g_i$. Does this yield $\sqrt{67.89}$? How many iterations were required? (This new algorithm is called *Newton's method*; by our count it requires the same number of steps, but more keystrokes, to key in data on a machine without memory.)

*P4. Follow Isaac Newton's reasoning to see that if you have an approximate root r_i for the equation $y^2 - x = 0$ and there is a small error δ_i in that approximation, $(r_i + \delta_i)^2 = x$, then

$$r_i{}^2 + 2\delta_i r_i + \delta_i{}^2 = x.$$

But if δ_i is a small number, much less than r_i, then $\delta_i{}^2$ is considerably less than the other two numbers $r_i{}^2$ and $2\delta_i r_i$ in this sum. Thus we may use the symbol $\doteq$ for approximate equality to say

$$r_i{}^2 + 2\delta_i r_i \doteq x$$

$$2\delta_i r_i \doteq x - r_i{}^2$$

$$\delta_i \doteq \frac{x - r_i{}^2}{2r_i}$$

$$r_i + \delta_i \doteq \frac{x - r_i{}^2 + 2r_i{}^2}{2r_i}$$

$$r_i + \delta_i \doteq \frac{x + r_i{}^2}{2r_i}.$$

Now this last formula is only approximate for $r_i + \delta_i = \sqrt{x}$. However, we may take $(x + r_i{}^2)/2r_i$ as an improvement over r_i, a better guess, which we will in turn call r_{i+1}:

$$r_{i+1} = \frac{x+r_i^2}{2r_i} \, .$$

Use this procedure to compute $\sqrt{67.89}$, with $r_0 = 8$. At which iteration did r_{i+1} first become equal to r_i? Is this method as easy to use on your machine as the method of our example? Did you have to reenter intermediate results?

P5. If for some reason we choose in finding $\sqrt{x}$ to guess $g = 0$ (perhaps x is small, say $x = \frac{1}{2}$), then we are presented with the iterative recipe

$$y_{i+1} = \frac{x}{y_i}$$

No matter what estimate we choose for y_0, we get $y_2 = y_0$, $y_1 = y_3$, and so forth, with no convergence. But it is true that if $y_0^2 < x$, then $y_1^2 > x$ (prove this!). Hence the average $\frac{1}{2}(y_0 + \frac{x}{y_0}) = (y_0^2 + x)/2y_0 = y_1$ may be a better estimate than y_0. This is, of course, Newton's formula, as we saw in Problem P4, but we have derived it in a different way. Draw a careful graph to illustrate the convergence of this method when $x = \frac{1}{2}$. Choose $y_0 = 3/4$ and make your graph magnify the region near the root of $y^2 - \frac{1}{2} = 0$. Then calculate $\sqrt{\frac{1}{2}}$. How many iterations were required? Show also that the method of the example works if the guess is $g = 3/4$.

The ancient Babylonians used one iteration of this recipe, in the disguise of: "If g_0 is a guess for $\sqrt{x}$, then let $x = g_0^2 + z$ and get a better guess $g_1 = g_0 + z/2g_0$." The Alexandrian Hero, or Heron, used this formula iteratively, just as Newton did 1700 years later, in a very early instance of our algorithmic methods.

P6. Discuss the "speed of convergence" of the algorithm in the Examples of this chapter. Begin your discussion by defining the nth error ε_n by $y_n = \varepsilon_n + \sqrt{x}$. Show by an arithmetic argument that as n gets larger and ε_n gets closer and closer to 0 ε_{n+1} gets close to $\frac{g-\sqrt{x}}{g+\sqrt{x}} \varepsilon_n$. This means that each successive error ε_{n+1} is smaller than

ε_n by a ratio that measures the error of g as a guess for $\sqrt{x}$. The better guess g we start with, the faster the convergence. (Compare Problem P3 in this light.)

Calculate the ratio $(g-\sqrt{x})/(g+\sqrt{x})$ for the example of $\sqrt{67.89}$, and then make a table of the ratios $\varepsilon_{n+1}/\varepsilon_n$ for each value of $n = 0, 1, 2, \ldots, 5$ to illustrate this theoretical argument.

P7. Describe at least one plausible situation in a field of your own current interest, perhaps biology or business or chemistry, where the techniques developed in this chapter for finding square roots may be applied to obtain a numerical solution that is useful. Discover such a real-life situation by surveying a current issue of an appropriate journal in your field. (See the Bibliography for some suggested journal titles.)

Answers to Starred Exercises and Problems

Exercises   1a. $y_4 = y_5 = 5.0990195$

                1b. $y_7 = y_8 = 5.9160798$

                1c. $y_4 = y_5 = 1.0535654$

                1d. $y_4 = y_5 = 12.247449$

                 3. $y_3 = y_4 = 5.9160798$

                 4. $r_1 = 1.6180340$ and $r_2 = -0.6180340$

                 8. $r_1 = 1.5815642$ and r_2 -1.3919328

               12. $\sqrt{\pi} = 1.7724539$

               13. $154g$

               14. $9\tfrac{1}{2}\%$

Problems   P4. $r_3 = r_4 = 8.2395388$

2

MORE FUNCTIONS AND GRAPHS

Introduction

We have seen that solutions for quadratic equations were known in
ancient times. But just 500 years ago the cubic equation was still
an enigma. Its final solution, developed at the University of
Bologna and published by Gerolamo Cardano in 1545, involves some
complicated algebra and the extraction of both square roots and cube
roots. The solution of the quartic equation is similar. As you will
see in this chapter, it is now possible to solve such equations on
the pocket calculator by simple and rapid methods.

As you study this chapter, your understanding of the concepts
of function and graphs will continue to grow. Working through the
various examples, exercises, and problems provided will help you to
improve your skills manipulating and evaluating functions and creat-
ing graphs. At the same time, as your competence with your calculator increases, you will be more and more able to concentrate on
the mathematics underlying your computations.

Chapter 2 provides a study of several iterative methods of solving equations of the form $f(x) = 0$, to find the zeros of $f(x)$. The most useful of these schemes is called the method of successive substitutions. We will apply this method to several cubic poly- nominals and to other functions. The roots that are calculated for $f(x) = 0$ will be displayed by graphs; they offer a new technique that aids graphing. In fact, the iterative methods themselves will be graphed to give a clear display of convergence. Synthetic division will be described and used in examples to factor out zeros.

In the Exercises we will illustrate the method of successive substitutions and compare it to several alternative algorithms. And in the Problems we will explore these methods theoretically, adding Newton's method for nth roots to the list.

DEFINITION: LIMITS OF SEQUENCES

We now define more clearly what is meant by convergence. If x_0, x_1, x_2, ... is an infinite sequence of numbers, we say that it *approaches a number* L or *has* L *as a limit* or *converges to* L if: no matter what degree of accuracy we require, there always exists a member x_n of the sequence so that x_n and also the subsequent members x_{n+i} for $i = 1,2,3,...$ all approximate L to within the given degree of ac- curacy. The sequences $y_0 = g$, y_1, y_2, ... of approximations for square roots that we calculated in Chapter 1 were convergent sequences in this sense.

EXAMPLE: $x^3 - 3x - 1 = 0$

Let us sketch a rough graph of the cubic polynomial function $f(x) = x^3 - 3x - 1$ (see Figure 2.1). When x is large, say $x \geq 2$, $f(x)$ is positive; and when x is large-negative, say $x \leq -2$, $f(x)$ is negative. Since $f(-2) = -3$ is negative and $f(-1) = 1$ is positive, the continuous graph of $f(x)$ must cross the x-axis somewhere. Thus there must be a zero z_1 for f *Figure 2.1* between -2 and -1. Similarly, $f(0) = -1 < 0$, $f(1) = -3 < 0$, and $f(2) = 1$, so there are two more zeros z_2 and z_3 with $-1 < z_2 < 0$ and $1 < z_3 < 2$. Since $f(x)$ is of degree

3, it has no more than 3 zeros, so these are all of them. We use a numerical trick to find z_2. Thus we may rewrite the equation as

$$x^3 - 3x - 1 = 0$$
$$x^3 - 3x = 1$$
$$x(x^2-3) = 1$$
$$x = \frac{1}{x^2-3}.$$

The last expression is true for every root x (we know that $\pm\sqrt{3}$ cannot be a solution to $f(x) = 0$ by the next-to-last equation above). Just as in the previous chapter, we now regard the two appearances of x in this last equation as two different numbers that are related by the equation.

We shall make a guess x_0 at the number z_2, say $x_0 = -0.5$, and calculate $x_1 = 1/(x_0{}^2 - 3)$, and in general $x_{i+1} = 1/(x_i{}^2 - 3)$, so that we obtain the results listed in Table 2.1. (Incidentally, the easiest method of evaluating the polynomial $x^3 - 3x - 1$ is as $(x^2-3)x - 1$.) Table 2.1 illustrates the same kind of convergence that we saw in Chapter 1 for the square root algorithm. The numbers x_i get closer and closer together, and the values $f(x_i)$ get closer and closer to 0. We have, for all practical purposes, found our zero z_2!

TABLE 2.1

	x	$f(x)$
x_0	-0.5	0.37
x_1	-0.3636364	0.0428249
x_2	-0.3487032	0.0037093
x_3		
x_4		
x_5		
x_6	-0.3472964	0.0000002
x_7	-0.3472964	0.[†]

[†]Notice that $x_6 = x_7$, although $f(x_6) \neq f(x_7)$. This is because our calculator is a 10-digit machine and these results are displayed as rounded off to 8 digits. Be sure to duplicate these computations on your own machine before continuing with the text.

16

This way of obtaining an algorithm is called the *method of successive substitutions*. You can understand how it works by considering the two functions $1/(x^2-3)$ and x, which are graphed in Figure 2.2.

The values of x for which these two functions are equal are exactly the zeros of $f(x)$. (Be sure you understand this fact. If you do not, think about the equations displayed above.) Figure 2.3 magnifies the region of Figure 2.2 near the intersection of the two graphs at z_2. Our iterative process corresponds graphically to following the horizontal and vertical dotted lines toward the intersection.

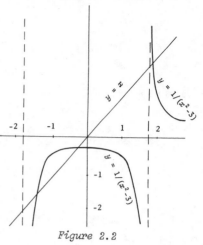

Figure 2.2

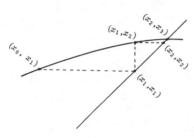

Figure 2.3

FINDING z_3 WITH ANOTHER ALGORITHM

We now attempt to use the same technique to find z_3, the zero between 1 and 2 for the function x^3-3x-1. For our first guess, we take $x_0 = 2$ and calculate as before (see Table 2.2).

TABLE 2.2

	x	$f(x)$
x_0	2.	1.
x_1	1.	-3.
x_2	-0.5	

Clearly something is amiss, and this process is not going to converge to z_3.[†] We go back to the equation $x(x^2-3) = 1$ and write

$$x^2-3 = 1/x$$

$$x^2 = 1/x + 3$$

$$x = \pm\sqrt{1/x + 3} \ .$$

Again we use the guess $x_0 = 2$ and calculate with our new algorithm
$x_{i+1} = +\sqrt{1/x_i + 3}$ to obtain the results shown in Table 2.3.

TABLE 2.3

	x	$f(x)$
x_0	2.	1.
x_1	1.8708287	-0.0645857
x_2	1.8800326	0.0049197
x_3	1.8793365	-0.0003703
x_4		
x_5		
x_6		
x_7		
x_8	1.8793852	0.

[†] A full explanation of this non-convergence requires the techniques
of Chapter 13 (see Problem P6, Ch. 13). However, the "trial and
error" method adopted here does obtain results and will suffice for
our present treatment of functions and graphs.

FINDING z_3 WITH SYNTHETIC DIVISION

Since $z_2 = -0.3472964$ is a zero of $f(x) = x^3 - 3x - 1$, the polynomial $f(x)$ is evenly divisible by $(x-z_2)$. That is, if we find a quotient polynomial $q(x)$ so that $f(x) = (x-z_2)q(x) + r$, then the remainder $r = 0$ (to see why this is so, substitute z_2 for x in the last equation). The algorithmic scheme for finding $q(x)$ is called *synthetic division*.

In general, we let $f(x) = f_0 x^n + f_1 x^{n-1} + \ldots + f_n$. The coefficients of the quotient polynomial $q(x) = q_0 x^{n-1} + q_1 x^{n-2} + \ldots + q_{n-1}$, such that $f(x) = (x-z)q(x) + r$ for some number z, are given by

$$q_0 = f_0$$

$$q_1 = q_0 z + f_1$$

$$\vdots$$

$$q_{i+1} = q_i z + f_{i+1}$$

$$\vdots$$

$$q_{n-1} = q_{n-2} z + f_{n-1}.$$

and
$$r = q_{n-1} z + f_n$$

These coefficients appear as intermediate steps in an evaluation of $f(x)$ at a number z. That is, the coefficients $q_0, q_1, \ldots, q_{n-1}$ are the contents of the successive pairs of parentheses in the expression $f(z) = (\ldots(((f_0)z + f_1)z + f_2)z + \ldots + f_{n-1})z + f_n$, and the whole expression $f(z)$ is $q_{n-1} z + f_n = r$.

In the case at hand, $f(x) = x^3 - 3x - 1 = (((1)x-0)x-3)x - 1$, and z is the number z_2 that is the zero for $f(x)$ that we have found: $f(z_2) = 0$. Hence $q_0 = 1$, $q_1 = z_2$ and $q_2 = z_2^2 - 3$, while $r = 0$. Thus synthetic division shows that $x^3 - 3x - 1 = (x-z_2)(x^2 + z_2 x + z_2^2 - 3)$, with zero for remainder.

19

The quadratic formula may now be used to solve the equation $q(x) = 0$ to find that

$$x = \frac{-z_2 \pm \sqrt{z_2{}^2 - 4(z_2{}^2 - 3)}}{2} = \frac{-z_2 \pm \sqrt{12 - 3z_2{}^2}}{2} \, ,$$

that is, $x = 1.8793852$ or $x = -1.5320889$. The first of these values is the zero z_3 for $f(x)$ that we have already found above. (Do you see why a zero of $q(x)$ must be a zero of $f(x)$ as well?) The other value for x above is the third zero for $f(x)$: $z_1 = -1.5320889$. Hence synthetic division offers an alternative method of finding the remaining zeros of a cubic polynomial once one zero is known.

EXAMPLE: $4x^3 + 3x^2 - 2x - 1 = 0$

We rewrite this equation as

$$4x^3 + 3x^2 - 2x = 1$$

$$x(4x^2 + 3x - 2) = 1$$

$$x = 1/(4x^2 + 3x - 2).$$

This defines the algorithm

$$x_{i+1} = (4x_i{}^2 + 3x_i - 2)^{-1},$$

which we begin blindly with a guess of $x_0 = 0$. The results are displayed in Table 2.4: $x_5 = x_6 = -0.3903882$. This number must be a zero of $4x^3 + 3x^2 - 2x - 1$.

TABLE 2.4

x_0	0.
x_1	-0.5
x_2	-0.4
x_3	-0.3906250
x_4	
x_5	-0.3903882
x_6	-0.3903882

Next we perform synthetic division to find that
$4x^3 + 3x^2 - 2x - 1 = (x - x_5) (q_0x^2 + q_1x + q_2) + r$ where

$$q_0 = 4$$

$$q_1 = 4x_5 + 3 = 1.4384472$$

$$q_2 = q_1x_5 - 2 = -2.5615528$$

$$r = q_2x_5 - 1 = 0.$$

The discriminant of the quadratic polynomial $q_0x^2 + q_1x + q_2$ is $\Delta = q_1^2 - 4q_0q_2$, which is positive. Accordingly, the quadratic formula can be used to find two real numbers that are zeros for this quadratic polynomial, and these numbers must be zeros for the above cubic polynomial as well. (See the Appendix for additional description of polynomial evaluation and synthetic division together with another detailed example.)

EXERCISES

1. In each of the cases below solve the given equation iteratively for a zero. Use the algorithm that is offered and begin with the indicated value of x_0.

 *a. $4x^3-7x-1 = 0$; $x_{i+1} = 1/(4x_i^2-7)$ with $x_0 = 0$

 *b. $x^3-5x^2-2 = 0$; $x_{i+1} = 5+2/x^2$ with $x_0 = 5$

21

*c. $x^4-9x-3 = 0$; $x_{i+1} = 3/(x^3-9)$ with $x_0 = 0$

d. $x^5-6x^2+1 = 0$; $x_{i+1} = 1/\sqrt{6-x^3}$ with $x_0 = 1$

2. Use synthetic division to divide $f(x) = x^3-3x-1$ by the linear factor $x-z_3$, where z_3 is the number determined in Table 2.3. That is, find a quotient polynomial $q(x)$ and a remainder r so that $f(x) = (x-z_3)q(x)+r$. Here r is a number, a polynomial of degree zero, and $q(x)$ is quadratic. Follow the example of synthetic division to find the coefficients of q and to show that $r = 0$.

Finally, use the quadratic formula (see Exercise 4, Ch. 1) to calculate the zeros of $q(x)$. Show by computation that these numbers are zeros of $f(x)$ as well.

3. Recalculate Table 2.3 and perform the missing steps to obtain z_3. In view of the failure of our earlier attempt to compute z_3 (in the discussion of the example), how can you be logically sure that your final result is really the number z_3, the largest zero of $f(x)$, accurate to the seventh decimal place?

4. Show that the method that begins with $x_1 = 1/(x_0^2-3)$ will not converge to z_1. Then find z_1 using the method that begins with $x_1 = -\sqrt{1/x_0+3}$. Do this by constructing a table just as was done above in the example.

5. Use the values given in the Examples for z_1, z_2, z_3 to sketch a large and accurate graph of $f(x) = x^3-3x-1$. Compute values of $f(x)$ in increments of 0.2 for $x = -2.2$, -2.0, -1.8, ..., 1.8, 2.0, 2.2. Label the special points z_1, z_2, z_3 and $f(0)$.

6. Find z_3 by the *method of interval-halving*. To begin, observe that $f(1) < 0$ and $f(2) > 0$ so that $1 < z_3 < 2$. Let $x_1 = 1.5$ and calculate $f(x_1) = -2.125 < 0$. Hence $1.5 < z_3 < 2$; let $x_2 = 1.75$ and continue this process 10 more steps, making a table as above. Compare this algorithm to that of Table 2.3 in the Example and Exercise 3, and discuss the differences.

7. Attempt to improve on the method of Exercise 6 above by estimating each number x_{i+1} yourself after inspection of $f(x_0)$, $f(x_1)$, ...,

22

$f(x_i)$. Did this speed the convergence very much? (This algorithm is called *bracketing*.)

*8. Let $f(x) = 4x^3+3x^2-1$. Show that $f(x) = 0$ if and only if $x = \sqrt{1/(4x+3)} = 1/\sqrt{4x+3} = (4x+3)^{-\frac{1}{2}}$. Since $f(0) < 0$ and $f(1) > 0$, there is a zero z for $f(x)$ between 0 and 1. Use the iterative recipe $x_{i+1} = (4x_i+3)^{-\frac{1}{2}}$, starting with $x_0 = \frac{1}{2}$, to find z. Next, use synethetic division to obtain a quadratic polynomial $q(x)$ such that $f(x) = q(x)(x-z)$. Then use the quadratic formula to calculate the other two zeros (if they exist) of $f(x)$. Check your answers.

9. Follow the scheme outlined in Exercise 8 to obtain a zero for $f(x) = \pi x^3+2\pi x^2-1$ between 0 and 1, using $x_0 = 0.2$ and $x_{i+1} = [\pi(x_i+2)]^{-\frac{1}{2}}$. Next use the quadratic formula to find any other roots of $f(x) = 0$ that exist. Finally, sketch a graph of f displaying your results.

*10. Attack the same problem as in Exercise 9, that of finding the zeros of $f(x) = \pi x^3+2\pi x^2-1$, in a different fashion. Notice that when $f(x) = 0$, then $\pi x^2(x+2) = 1$, or $x = 1/\pi x^2 - 2$. The facts that $f(-2) = -1$ and $f(-1) = \pi - 1 > 0$, plus the appearance of the factor $x + 2$ above suggest that there is a zero for $f(x)$ near -2. Try $x_0 = -2$ and $x_{i+1} = (\pi x_i^2)^{-1} - 2$. When this process converges to a zero z, divide $f(x)$ by $x-z$ just as in the two previous exercises and search for further zeros in the quadratic factor of f. Finally, sketch a graph of f displaying your results.

11. A certain drug is found to raise human body temperature according to the formula $T(D) = 1.81 D^2 - D^3/3$. Here D is the dosage in grams in the range $0 \leq D \leq 3.5$, and $T(D)$ is the change in body temperature in degrees Fahrenheit due to that dosage (when there is no trace of the drug in the body to begin). Find the dosage (to the nearest milligram) required to raise body temperature by $5°$.

12. A button manufacturer finds that the plastic raw material for a certain item in her line costs $7.42 per thousand finished buttons. The machine that makes these buttons costs $30 to set up for a run and then $100 \sqrt[3]{x}$ dollars labor cost to run x thousand buttons. Office work to handle one order costs $20. If the selling price is $10 per thousand, how large must an order be for the manufacturer to realize

23

20% of her billing as profit? (Hint: the equation for x is $7.42x + 30 + 100\sqrt[3]{x} + 20 = 8x$. Define a new variable $y = \sqrt[3]{x}$, rewrite the equation in terms of y, and solve for y.)

PROBLEMS

*P1. Test the usefulness of the formula $x_{i+1} = (x_i^3-1)/3$ for finding the three zeros of $f(x) = x^3-3x-1$. Then test the opposite formula, $x_{i+1} = \sqrt[3]{3x_i+1}$, for its convergence near each zero of $f(x)$. Sketch graphs illustrating the convergence of each of these algorithms and also illustrate their non-convergence at the zeros where that non-convergence occurs. (Compare Figures 2.2 and 2.3.)

P2. Find necessary and sufficient conditions on the linear function $g(x) = mx + b$, where m is the slope of the line and b is its y-intercept, in order that the algorithm $x_{i+1} = mx_i+b$ will work to find the intersection of the graph of $y = g(x)$ with the graph of $y = x$. The hint for this problem is the graphic display by dotted lines of the convergence. You need only ask yourself what the dotted lines mean and where they begin and end. You should then be able to find the algebraic requirement that will insure convergence.

*P3. Let x be an approximation to the nth root of the number y, so that there is some small error δ for which $x + \delta = \sqrt[n]{y}$ or $(x + \delta)^n = y$. Following Isaac Newton's reasoning, expand $(x + \delta)^n$ by the binomial theorem to get

$$(x+\delta)^n = x^n + nx^{n-1}\delta + \frac{n(n-1)}{2} x^{n-2}\delta^2 + \ldots + \delta^n.$$

Notice here that all the summands after the first two terms on the right hand side are multiples of δ^2. But δ was agreed to be very near to zero, so δ^2 is very small indeed, so small that you will make an insignificant error if you neglect all the terms that are multiples of δ^2 and write (with $\doteq$ meaning approximate equality)

$$y = (x+\delta)^n \doteq x^n + nx^{n-1}\delta$$

$$nx^{n-1}\delta \doteq y - x^n$$

$$\delta \doteq (y - x^n)/nx^{n-1}$$

$$x+\delta \doteq \frac{y-x^n}{nx^{n-1}} + x = \frac{y-x^n+nx^n}{nx^{n-1}} = \frac{y+(n-1)x^n}{nx^{n-1}} \ .$$

In the last form the approximate value for $x + \delta$ should be a better estimate for $\sqrt[n]{y}$ than is x, the estimate you started with. Accordingly, use

$$x_{i+1} = \frac{y+(n-1)x_i^n}{nx_i^{n-1}} \ ,$$

together with a reasonable starting guess x_0, to find $\sqrt[2]{67.89}$, $\sqrt[3]{67.89}$, $\sqrt[4]{67.89}$. Do you think it was easier to go through one iterative process on your machine to find $\sqrt[6]{67.89}$, or would it be better to find $\sqrt[3]{\sqrt[2]{67.89}}$ or $\sqrt{\sqrt[3]{67.89}}$?

*P4. Show that the method of successive substitutions may be applied to the equation $x^2 - 67.89 = (x^2 - 64) - 3.89 = 0$ to obtain the algorithm described in Chapter 1.

Follow the same line of reasoning to solve $x^2 - 3x + 1 = 0$ by first making a guess $g = 1/3$, then factoring $x^2 - 3x + 1 = (x-g)(x-h) + r$. Finally, use the last expression to define an algorithm for successive substitutions. Pursue your algorithm to a solution for $x^2 - 3x + 1 = 0$ and check your work. How does this method compare with a use of the quadratic formula?

P5. Picture two ladders of lengths 4m and 5m, each leaning directly across an alley of width 3m. If the ladders are propped against opposite walls of the alley so that they cross side by side, how high is the crossing point?

*P6. Solve $x^3 - 4x^2 + 2x - 7 = 0$ by the method of successive substitutions.

Answers to Starred Exercises and Problems

Exercises   1a. $x_5 = x_6 = -0.1445843$

             1b. $x_4 = x_5 = 5.0775742$

             1c. $x_4 = x_5 = -0.3319837$

              8. $z = 0.4554100$ is the only zero.

           10. $x_7 = x_8 = -1.9130218$

Problems   P1.   The first recipe finds z_2 only; the second finds z_1 and z_3 but not z_2.

            P3. $(67.89)^{1/3} = 4.0794530$

                  $(67.89)^{1/6} = 2.0197656$

                  $(67.89)^{1/7} = 1.8267812$

            P4. $z = 0.3819660$

            P6. $z = 3.9430114$ is the only zero.

3

LIMITS AND CONTINUITY

INTRODUCTION

We have been assuming up to now that the functions we were working
with were "continuous." That is, we have assumed that if $f(r) = 0$
and x_0, x_1, x_2, ... were numbers that got closer and closer to r,
then the numbers $f(x_0)$, $f(x_1)$, $f(x_2)$, ... would get closer and closer
to $f(r) = 0$. More generally, a function f is *continuous* if for each
point y and each sequence x_0, x_1, x_2, ... = $\{x_i\}$ that has y as limit,
$x_i \to y$, we have $f(x_i) \to f(y)$. We may also express this by writing

$$\lim_{x \to y} f(x) = f(y)$$

Although the use of limiting methods stretches back to Archimedes
(287?-212 B.C.), rigorous definitions were only offered beginning with
Bolzano and Cauchy in the nineteenth century A.D. We shall now examine this operation of "taking limits" more thoroughly. Our study

27

will illustrate theorems about limits of sums, products, and quotients
of functions by means of detailed numerical Examples. This numerical
work will display tabulations of converging sequences. The Examples
will illustrate the convergence of a sequence of values for a function by tabulating the successive members of the sequence. Exercises
continue these displays and evaluate some limits that are not at all
obvious. In the section of Problems you will be able to study some
restrictions on numerical limit taking, examine limits for some
exponential functions, and define the limit as the variable tends
toward infinity.

EXAMPLE: $f(x) = 3x + 4$

Let us calculate the values of $f(x) = 3x+4$ for values of x near 2.
Say $x_0 = 2+1$, $x_1 = 2+0.1$, ..., $x_i = 2+10^{-i}$: $f(x_5) = 3(2+10^{-5})+4 =$
$3\text{X}2+4+3(0.00001) = f(2)+0.00003$, and so forth. We say $\lim\limits_{x\to 2} 3x+4 = 10$.
Now let

$$g(x) = \frac{3\pi x}{2} + 2\pi = \frac{\pi}{2}(3x+4).$$

What is $\lim\limits_{x\to 2} g(x)$? This function $g(x)$ is $\pi/2\, f(x)$, so

$$\lim\limits_{x\to 2} \frac{\pi}{2} f(x) = \frac{\pi}{2} f(2) = 5\pi.$$

Just for fun, we check the values of $g(x)$ (using $\pi = 3.1415927$, so
$g(2) = 15.707963$). Table 3.1 presents the values of x greater than
2, and Table 3.2 lists values of x less than 2.

28

TABLE 3.1[†]

x	$g(x)$
3.	20.420352
2.1	16.179202
2.01	
2.001	15.712676
2.0001	
2.00001	
2.000001	15.707968
2.0000001	15.707964

TABLE 3.2

x	$g(x)$
1.999	15.703251
1.9999	
1.99999	
1.999999	15.707959
1.9999999	15.707963

This computation illustrates the following THEOREM: *if* $\lim\limits_{x \to y} f(x) = k$ *and* c *is a number then* $\lim\limits_{x \to y} cf(x) = ck$. It's easy to see that $\lim\limits_{x \to y}\left(f(x) + c\right) = k + c$, too.

[†]Remember to read with your machine *ON* so that you can duplicate these results for yourself as well as fill in the blanks in the tables.

These rules together imply that for our function $f(x) = 3x+4$, $\lim_{x \to 2} f(x)-10 = 0$, and thus that for each positive integer n, $\lim_{x \to 2} n! \, (f(x)-10) = 0^{\dagger}$. We tabulate two sequences of results, one for $n = 4$ and one for $n = 8$ in Table 3.3.

TABLE 3.3

x	$4! \; (f(x)-10)$	$8! \; (f(x)-10)$
2.1	7.2	12096.
2.001	0.072	120.96
2.00001		
2.0000001	0.0000072	0.012096
2.	0.	0.
1.9999999	-0.0000072	-0.012096
·1.99999	-0.00072	-1.2096

If we interpret this table using the same standards we have used previously, it is not clear that there is convergence to 0. The function $h(x) = 8!(f(x)-10)$ has values that never get as close as 1/100 to 0, whereas we have been obtaining convergence in at least 7 digits. The fact that $h(2) = 0$ exactly is reassuring, but that sort of fact will not be available on an 8-digit machine when the limiting value for x is not an 8-digit rational number (see Exercise 2, for instance, when $y = \sqrt{2}$ is irrational).

This is a numerical puzzle, and there is no simple answer to it. An 8-digit machine is a kind of microscope with which to examine the behavior of a function at a given point, and its magnification is 100 million-fold. But even with this enormous magnification, there are some things that are still fuzzy, and other details that remain entirely invisible. Thoughtful experience with numerical examples is the best guide.

$^{\dagger}n!$, read "n-$factorial$," is the product of the positive integers up to n: $1!=1$, $2!=2$, $3!=6$, $4!=24$, ..., and $n!=n \times (n-1)!$ By convention, $0!=1$.

EXAMPLES: THEOREMS FOR SUMS AND PRODUCTS

$$\lim_{x \to y} (f(x)+g(x)) = \lim_{x \to y} f(x) + \lim_{x \to y} g(x),$$

if the latter two limits exist (see Exercise 3).
Similarly

$$\lim_{x \to y} (f(x)g(x)) = \left(\lim_{x \to y} f(x)\right) \left(\lim_{x \to y} g(x)\right).$$

If we use the function $g(x) = 3\pi x/2 + 2\pi$ from above and let $f(x) = (x-1)$, then

$$f(x)g(x) = \left(\frac{3\pi x}{2} + 2\pi\right)(x-1) = \frac{3\pi x^2}{2} + \frac{\pi x}{2} - 2\pi .$$

Clearly $\lim_{x \to 2} f(x) = 1$, so $\lim_{x \to 2} f(x)g(x) = 5\pi$. We calculate this in
in Table 3.4

TABLE 3.4

x	$f(x)\ g(x)$
3.	40.840705
2.1	17.797122
2.01	15.912638
2.0012345	
2.0001987	
2.0000164	
2.0000015	
2.0000001	15.707965
2.	15.707963
1.9999999	15.707961
1.9999997	
1.99999	15.707759

31

This table should be compared with Tables 3.1 and 3.2. Convergence of $f(x)g(x)$ in Table 3.4 is slightly slower but quite similar to that of $g(x)$ itself in the earlier tables.

EXAMPLES: LIMITS OF QUOTIENTS

For limits of quotients $f(x)/g(x)$ you would expect the same sort of rule to be obeyed as for products

$$\lim_{x \to y} \frac{f(x)}{g(x)} = \frac{\lim\limits_{x \to y} f(x)}{\lim\limits_{x \to y} g(x)}.$$

Indeed, this is a THEOREM, *provided that the limits exist for both f(x) and g(x) when x approaches y and also provided* $\lim\limits_{x \to y} g(x) \neq 0$. In the latter case, when $\lim\limits_{x \to y} g(x) = 0$, the above rule makes no sense

TABLE 3.5

x	$(x-1)$ / $(\sqrt{x}-1)$
2.	2.4142136
1.1	2.0488089
1.01	
1.001	
1.0001	
1.00001	
1.000001	2.
.	.
.	.
.	.
0.999999	2.
0.99999	
0.9999	
0.999	
0.99	1.9949874

32

since division by 0 is not possible. Nevertheless, $f(x)/g(x)$ may still have a limit when $g(x)$ has limit 0. For instance, $\lim\limits_{x\to1} \frac{x-1}{x-1} = 1$

$\lim\limits_{x\to1} \frac{(x-1)^2}{x-1} = \lim\limits_{x\to1} x-1 = 0$. A more interesting example is $\lim\limits_{x\to1} \frac{x-1}{\sqrt{x}-1}$.

It is clear from Table 3.5 that this limit is 2. Here our calculator was of genuine help.

EXERCISES

1. Evaluate the following limits:

*a. $\lim\limits_{x\to0}(\sqrt{1+x} - 1)/x$

*c. $\lim\limits_{y\to0}(\sqrt{y+9} - 3)/y$

*b. $\lim\limits_{x\to5}(x-5)/(\sqrt{5x} - 5)$

d. $\lim\limits_{t\to0}(\sqrt{7t+49} - 7)/t$

2. Calculate $\lim\limits_{x\to\sqrt{2}} 4!(x^2-2)$, and $\lim\limits_{x\to\sqrt{2}} 8!(x^2-2)$, compiling values of the functions as in Table 3.3. How much confidence would you place in your limit values? Why?

3. Find the limits as $x\to2$ of the two functions $f(x) = x^2+3x+2$ and $g(x) = 3x/(1-x)$ and then $\lim\limits_{x\to2}(f(x)+g(x)+6)$. Do this by constructing a column of values for each function corresponding to a column of argument values x as in Tables 3.1 and 3.2.

4. Compile data like that of Table 3.4 for the quotient function $g(x)/f(x)$ to find $\lim\limits_{x\to2} \frac{3\pi x/2 + 2\pi}{x-1}$. Compare your table with Tables 3.4 and 3.1 and 3.2 as to the speed of convergence.

5. Make a table of values to find $\lim\limits_{x\to3} \frac{\sqrt{x} - \sqrt{3}}{x-3}$.

*6. Make a table of values to find $\lim\limits_{x\to1} \frac{2x^2 - (3x+1)\sqrt{x}+2}{x - 1}$.

*7. The function $\sqrt{1-x^2}$ is not defined for values of x greater than 1; nevertheless we may be interested in its limit as x nears 1. We denote by $\lim\limits_{x\uparrow1} \sqrt{1-x^2}$ by the limit of this function as x approaches 1 by taking on only values less than 1. Similarly, $\lim\limits_{x\downarrow1} \sqrt{x^2-1}$ means the *one-sided limit* as x approaches 1 from above. These types of

limits correspond to Tables 3.2 and 3.1, respectively. Make a similar table to investigate $\lim\limits_{x \uparrow 1} \dfrac{1-x}{\sqrt{1-x^2}}$.

8. Make a table to establish the following limit:

$$\lim_{x \to -2} \frac{x^4 - x^3 - 24}{x^2 + x - 2} \ .$$

9. If a bank deposit of $100 earns 6% interest for a year, then it
returns $106 at the end of the year. But if 3% interest is added
after 6 months, and then the new balance of $103 earns 3% interest
over the last half year, then at year's end there is a return of
103 X 1.03 = 106.09 dollars. The extra 9 cents is the interest paid
during the last half year on the $3 interest that was paid on the
$100 principal during the first half year.

In the first case above, simple interest was paid. In the
second case, interest was *compounded* semiannually at the annual rate
of 6%, to yield 6.09% per year. Calculate the yearly yield if 6%
interest is compounded quarterly (every 3 months)? Compounded daily?
Do you think the yield could be made arbitrarily large this way, or
is there a limit to this numerical process? That is, is there a
yearly rate corresponding to "continuous" compounding of 6% interest?
(This topic is treated fully by Example, Exercises, and Problems in
Chapter 8.)

PROBLEMS

P1. We have used powers of 10 heavily in our computations of limits;
that is, in each table of this chapter the values of the argument
that we used differ from the limiting value by powers of 10 that
diminish toward 0. But could an unscrupulous function, knowing of
this habit of ours, deceive us by having special values just as
these points we have chosen?

Discuss this question by describing such a function (sketch its
graph, perhaps with a second graph showing magnification by change
of scale near the limit point). Would you recognize such a function
from its recipe? What could go wrong with our calculation of its

limit? Could the limit be a different number than the one we calculate? Could we avoid this kind of problem by using powers of 2 in
place of powers of 10?

Make a table of values for $f(x) = 987654y^3-32y-1$ at $x_0 = 1/32$
and $x_{i+1} = 1/(987654x_i^2-32)$. Can you find the zero(s) for $f(x)$?

P2. We know what a^n means for an integer n and a positive number a
and that $a^{-n} = 1/a^n$. We also know about $a^{1/n} = \sqrt[n]{a}$, and we can com-
bine these operations to find $a^{m/n} = \sqrt[n]{a^m} = \left(\sqrt[n]{a}\right)^m$. This expression
represents the positive number that when raised to the nth power
gives a^m. Hence "a^x" makes sense when x is a positive rational
number; we now assume that we have extended this function to be
defined for all positive real $x^{\dagger}$. Investigate the behavior of this
function near 0 by computing a table of values a^{x_i} for $a = 67.89$ and
and $x_0 = 1$, $x_i = x_{i-1}/2$. This is, of course, done by repeatedly
taking square roots. What is $\lim_{x \downarrow 0} (67.89)^x$?

Make another similar table for $\lim_{x \downarrow 0} (0.0000123)^x$.

Can you draw a general conclusion from your results?

P3. The limit of $f(x)$ as x tends toward infinity, $\lim_{x \to \infty} f(x)$, is
defined to be the number $\lim_{x \downarrow 0} f(1/x)$ if that limit exists (see Exercise 7). For example, if $f(x) = 1/x$, then $\lim_{x \to \infty} \frac{1}{x} = \lim_{x \downarrow 0} x = 0$.
Similarly

$$\lim_{x \to \infty} \frac{2x^2+3x+4}{x^2-5x-6} = \lim_{x \downarrow 0} \frac{2/x^2+3/x+4}{1/x^2-5/x-6}$$

$$= \lim_{x \downarrow 0} \frac{2+3x+4x^2}{1-5x-6x^2}$$

$$= \frac{2}{1}$$

$$= 2.$$

---

†This extension really is not necessary for our purposes since our
calculating machines deal only with rational numbers.

Verify by substituting increasing values 1, 10, 100, 1000, 10000 for x directly into the expression $\dfrac{2x^2+3x+4}{x^2-5x-6}$ and tabulating the results that this limit is indeed approached by the values of this function as the argument x gets large without limit.

Next investigate the limit of $x^8/2^x$ as x tends toward infinity by tabulating its values at x = 1, 10, 20, 30, 40, 50, 60, 70, 80 (calculate $\dfrac{80^8}{2^{80}} = \left(\dfrac{80}{2^{10}}\right)^8$, etc., to avoid machine overflow).

P4. Calculate the number $e = \lim\limits_{x\to\infty} (1+1/x)^x$ by tabulating values for x = 1, 10, 100, 1000, ..., taking enough values of x to see the values of the function repeat themselves.

Next, calculate in the same fashion $\lim\limits_{x\to\infty} (1+2/x)^x$ and show that your limit is e^2. Make a theoretical proof of this fact, given that the first limit was e.

*P5. Make a table to establish the limit (note Problem 2)

$$\lim_{x\to 0} \frac{67.89^x-1}{x}$$

If your machine has a memory, calculations will be shortened if you store 67.89^{x_i} while computing $\dfrac{67.89^{x_i} - 1}{x_i}$, then let $x_{i+1} = x_i/2$ and find $67.89^{x_i+1} = \sqrt{67.89^{x_i}}$. Computational error enters this calculation more rapidly than convergence occurs (can you see why?), so expect only 2- or 3-figure accuracy.

P6. Give a reasoned argument that the result of Exercise 1a implies that $x/2 + 1$ is a good approximation for $\sqrt{1 + x}$ when x is small enough. Then illustrate your theorem by calculating this approximation and its error for x_0 = 0.123, x_1 = 0.00234, and x_2 = 0.0000567.

Answers to Starred Exercises and Problems

Exercises	1a.	½	7.	0	Problems	P5.	4.2178887
	1b.	2					
	1c.	1/6					
	6.	-1					

4

DIFFERENTIATION, DERIVATIVES, AND DIFFERENTIALS

INTRODUCTION

We shall now meet one of the subtlest and most beautiful concepts the human mind has yet created, the derivative. Since Sir Isaac Newton (1642-1727) and Baron Gottfried Wilhelm von Leibniz (1646-1716) first taught this idea, it has given us immeasurably valuable insight into change and the way our universe unfolds in time. It has been used to predict the future configurations of stars and planets, moving rocks and rockets, the stock market and housing costs, bacterial growth and radioactive decay.

Our study will use Examples to gain numerical and geometric insight. At first we shall need to work out some simple arithmetic rules to ease calculations. Then applications will be explored in the Exercises and in some Problems. One Problem studies the theory further, showing how the error in the differential approximation goes to zero faster than Δx. Another Problem defines second derivatives, and a third constructs Newton's method for finding zeros for functions.

37

EXAMPLE: $f(x) = x^2$

The *derivative* of a function f at a point a is defined to be

$$f'(a) = \lim_{x \to a} \frac{f(x) - f(a)}{x - a}$$

if that limit exists. If we calculate that limit for $f(x) = x^2$ and $a = 1.2$, with $x_i = 1.2 + 10^{-i}$, we get the results shown in Table 4.1. These results are illustrated in Figure 4.1.

Table 4.1

x	$\dfrac{f(x) - f(1.2)}{x - 1.2}$
1.3	2.5
1.21	2.41
1.201	
1.2001	
1.20001	2.4
...	...
1.19999	2.4
1.1999	
1.199	2.399

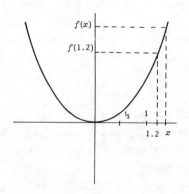

Figure 4.1

38

There is another, equivalent way of expressing the derivative as a limit:

$$f'(a) = \lim_{h \to 0} \frac{f(a + h) - f(a)}{h} .$$

The equivalence of these two limits may be seen by setting $x = a + h$ or $x - a = h$, so that h represents a small change or *increment* in the values of x from $x = a$. It is then the same to consider the limit as x approaches a or the limit as h approaches 0.

At each stage in this limiting process, the *difference quotient*

$$\frac{f(x) - f(a)}{x - a} \quad \text{or} \quad \frac{f(a + h) - f(a)}{h}$$

may be visualized as the ratio of lengths of two intervals. The denominator is the (signed) length of a small interval $[a,x]$ or $[a, a + h]$ of arguments, while the numerator is the (signed) length of the small interval $[f(a), f(x)]$ of values of the function f. That is, the difference quotient is the ratio by which f "stretches" or "magnifies" the interval $[a,x]$. The limiting value $f'(a)$ is, then, the ratio of stretching or magnifying that f effects right at the point a. The derivative at a may also be thought of as the rate of change at a in the values of the function.

You will recall that the *slope* of a line is its rate of climb to the right or the amount by which it rises (a fall is counted as a negative rise) as you go one horizontal unit to the right.

EXAMPLE: $f(x) = 1/x$

For instance, let $f(x) = 1/x$ and $a = 1.2$ as in Figure 4.2. If we go horizontally from 1.2 to x, a change or increment of $h = x-1.2$ units, the chord "rises" from $1/1.2$ to $1/x$ for a total rise of $1/x-1/1.2$. The slope of this line is thus $\dfrac{1/x-1/1.2}{x-1.2}$. If we magnify a portion of this graph (see Figure 4.3), we can see the limiting process: the derivative $f'(1.2)$ is the limit of the slopes of these chords as $x \to a$. We calculate this slope in Table 4.2.

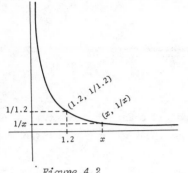

Figure 4.2   Figure 4.3

TABLE 4.2

x	$\dfrac{1/x-1/a}{x-a}=\dfrac{-1}{ax}$
1.21	-0.6887052
1.2001	
1.200001	
1.2000001	-0.6944444
...	...
1.1999999	-0.6944445
1.199999	
1.1999	-0.6945023

The limiting result is $-(1.2)^{-2}$, which is negative since $f(x) = 1/x$
is decreasing in its values as x increases past 1.2.

RULES FOR DIFFERENTIATION

The simple theorems we saw in the last chapter about sums, products,
and quotients of limits readily yield THEOREMS ABOUT DERIVATIVES:
If f *and* g *are functions that have derivatives at* a, *and* b *is a real*
number, then

$$(f+g)'(a) = f'(a) + g'(a)$$
$$(bf)'(a) = bf'(a)$$
$$(fg)'(a) = f'(a)g(a) + f(a)g'(a)$$
$$(f/g)'(a) = \frac{g(a)f'(a) - f(a)g'(a)}{g(a)^2}$$

40

where the last rule, for quotients, only makes sense when $g(a) \neq 0$. Also, we see from the definition of derivative that if $f(x) = b$ is a constant function, then $f'(a) = 0$ for every number a. Similarly if $g(x) = x$, then $g'(a) = \lim\limits_{x \to a} \dfrac{x - a}{x - a} = 1$ for every a.

DERIVATIVES FOR POLYNOMIALS

It is easy to show from these rules that $(x^2)' = 2x$, $(x^3)' = 3x^2$, and (by induction) $(x^n)' = n\, x^{n-1}$. Hence the derivative of any polynomial

$$p(x) = a_0 x^n + a_1 x^{n-1} + \ldots + a_{n-1} x + a_n$$

at any point $a = x$ may be written down immediately. The derivative of the sum is the sum of the derivatives of the summands, so

$$p'(x) = n\, a_0 x^{n-1} + (n-1)a_1 x^{n-2} + \ldots + a_{n-1}$$

at every point x. This defines a new function $p'(x)$ by giving its values for each x. Thus the derivative of a polynomial function of degree n is always a polynomial function of degree $n-1$.

This line of reasoning has relieved us of calculating limits in order to find the derivatives for polynomials or even quotients of polynomials (a quotient of polynomials is called a *rational function*).

EXAMPLE: THE DERIVATIVE OF $\sqrt{x}$

The product rule for differentiation may be used to find the derivative of $f(x) = \sqrt{x}$. Let $g(x) = x$ for every number x; then $f(x)f(x) = [f(x)]^2 = g(x)$. By the product rule $g'(x) = f'(x)f(x) + f(x)f'(x) = 2f(x)f'(x)$. The derivative of $g(x)$ was found above to be $g'(x) = 1$. Hence $2f(x)f'(x) = 1$ and $f'(x) = 1/2f(x)$. That is, the derivative function of the square root function $f(x)$ is $1/2\sqrt{x}$. A frequently used notation for this fact is $\dfrac{d}{dx} \sqrt{x} = \dfrac{1}{2\sqrt{x}}$.

DIFFERENTIALS

There is a way of thinking that uses these limiting values backwards

41

so that we can estimate the value of the difference quotient
$\frac{f(x) - f(a)}{x - a}$ as being nearly equal to its limit $f'(a)$ whenever x is
close to a. This enables us to estimate the change $f(x) - f(a)$ in
values of the function f when we change its argument by a small
amount $x-a$ near a:

$$f(x) - f(a) \doteq (x-a)f'(a).$$

Here the dot above the equals sign indicates *approximate equality*.
That is, $\doteq$ means that the two numbers are nearly equal, though they
do not necessarily agree to seven decimal places. We continue to
use equality to mean agreement in at least seven decimal places;
that is, agreement on an 8-digit machine.

The above approximation is so useful that it has a special name:
the *differential* df of a function f at a point $x = a$ is $df = f'(a)dx$.
Here dx represents a change or increment in the value of x (see Figure 4.4). The above approximation then says that the differential

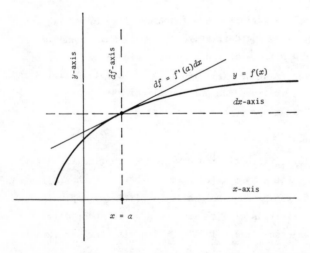

Figure 4.4

is nearly equal to the change in value for f:

$$df = f'(a)dx \doteq f(x) - f(a) .$$

42

Hence we may estimate the new value $f(x)$ of f as

$$f(x) \doteq f(a) + f'(a) \; dx \; .$$

EXAMPLE: $\sqrt{103}$

Let us use the above reasoning to estimate $\sqrt{103}$ as follows. Let $f(x) = \sqrt{x}$ and $a = 100$, so that $dx = 3$. Then $f'(a) = 1/2\sqrt{100} = 1/20$ and we estimate

$$\sqrt{103} \doteq 10 + 3/20 = 10.15 \; .$$

In fact, $\sqrt{103} = 10.148892$, so this estimate is off by about 1/1000.

EXAMPLE: $\sqrt{142.3}$

Here $f(x) = \sqrt{x}$ again, and we take $a = 144$, so $\sqrt{a} = 12$, and $f'(a) = 1/2\sqrt{144} = 1/24$. The increment $dx = 142.3 - 144 = -1.7$ is negative, so

$$\sqrt{142.3} \doteq 12 - 1.7/24 = 11.929167 \; .$$

This time the error in the estimate is 0.0002.

EXAMPLE: PAINTING A CUBE

Suppose we wish to estimate the volume of a paint film 0.012 inches thick on a metal cube with edges of length 3.456 inches. The cube has volume $V(x) = x^3 = (3.456)^3$. The differential $dV = 3x^2 dx$, the derivative $3x^2 = 3 \times (3.456)^2 = 35.831808$, and the increment $dx = 0.024$ (the edge measurement changes by twice the film thickness). We calculate that $dV = 0.8599634$. In this example it is easy to check the exact value as $(3.456 + 0.024)^3 - 3.456^3 = 0.8659492$. The error is less than 1%, which is accurate enough for many purposes. If the film had been an electroplated layer of gold 3 millionths of an inch thick, $dx = 0.000006$, we could use our already computed value $V'(3.456) = 35.831808$ to calculate $dV = 0.0002150$. Again we check our result: the correct volume of gold is $3.456006^3 - 3.456^3 = 0.0002150$; thus this use of the differential gives excellent accuracy.

43

Incidentally, if you calculate the paint thickness the easy way, you calculate the area of a cube face as x^2 and count 6 faces, each having x^2 times film thickness in added volume (see Figure 4.5).

This is exactly the method of differentials. It is in error because it ignores the part of the film at the edges and corners that is not straight out from any face. It is surprising, isn't it, that there is nearly 1% error when this simple method is applied to

Figure 4.5

the paint film? (In fact, a paint film would not be sharp at the edges, so the error is not quite that large.)

The differential for f at a may be pictured as approximating a change Δf in the values of f by the corresponding change along the line tangent to the graph of f at the point a, as in Figure 4.6.

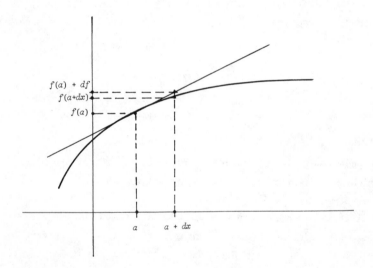

Figure 4.6

COMPOSITES AND INVERSES

If $f(x)$ and $g(x)$ are functions, the *composite function* or *composition* f ∘ g(x) is the function whose values are $f \circ g(x) = f\big[g(x)\big]$. For instance, if $f(x) = \sqrt{x} - 3$ and $g(x) = 2x + 1$, then $f \circ g(x) = \sqrt{2x + 1} - 3$. The *chain rule* for differentiating composite functions

44

says that *if* f(x) *and* g(x) *have derivatives* f'(x) *and* g'(x) *then*

$$(f \circ g)'(x) = f'[g(x)] \times g'(x).$$

This is easy to visualize (see Figure 4.7) as a stretching or magnification. The combined stretch caused by the composite function $f \circ g$ at x is the stretching $g'(x)$ that g effects in going from x

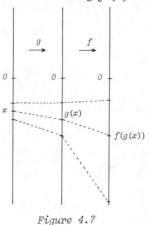

to $g(x)$ multiplied by the effect $f'(g(x))$ of f in going from $g(x)$ to $f[g(x)]$. A special case of this rule is the square root function $g(x) = \sqrt{x}$, which composes with the squaring function $f(x) = x^2$ to give $f \circ g(x) = x$ when $x \geq 0$. Thus

Figure 4.7

$x' = 1 = (f \circ g)'(x) = 2g(x)g'(x) = 2\sqrt{x}g'(x)$, giving $g'(x) = 1/2\sqrt{x}$, as we proved above by the product rule.

The squaring function and the square-root function are said to be *inverse functions* for each other where both are defined on the non-negative numbers. The chain rule gives us a method of finding the derivative of a function when we know the derivative of its inverse function. Thus the nth root function $g(x) = \sqrt[n]{x} = x^{1/n}$ is inverse to the nth power function $f(x) = x^n$, and $1 = ng(x)^{n-1}g'(x) = nx^{\frac{n-1}{n}}g'(x)$. Hence

$$g'(x) = \frac{1}{\frac{n-1}{nx^{\frac{n-1}{n}}}} = \frac{1}{nx^{1-1/n}} = \frac{1}{n}x^{1/n-1}$$

is the derivative of $\sqrt[n]{x} = x^{1/n}$.

EXERCISES

1. Find the value of the differential df in each case below:

*a. $f(x) = 3$, $a = 3$, $dx = 0.1$ *d. $f(x) = 5x^8$, $a = \frac{1}{2}$, $dx = 0.1$

*b. $f(x) = \sqrt{x}/3$, $a = 2$, $dx = 0.2$ e. $f(x) = x^{3/2}$, $a = 1/3$, $dx = 0.01$

*c. $f(x) = x^2 - 7x$, $a = 1$, $dx = 1$ f. $f(x) = 3/\sqrt{x}$, $a = 2$, $dx = 0.23$

2. Argue another derivation for the derivative of $f(x) = 1/x$ by observing that f is its own inverse function, $f[f(x)] = \dfrac{1}{1/x} = x$.

Then use its differential to estimate $1/x$ when $x = 1.1$, 1.001, 1.00001, 0.9, 0.999, 0.99999. Also use the differential to estimate $1/x$ if $x = 499$, 500.01, 500.0001. For each estimate, calculate the correct value and the error in your estimate.

3. Use the differential of $\sqrt{x}$ to estimate $\sqrt{67.89}$, $\sqrt{35}$, $\sqrt{35.99}$, $\sqrt{36.00001}$, $\sqrt{50}$, $\sqrt{4899}$. For each estimate, calculate the correct value and the error in your estimate.

4. A rocket is fired straight upward so that, until burnout at $t = 300$ seconds, its height $h(t)$ in meters above the earth at t seconds after launch is given by

$$h(t) = 0.03t^3 + 67.89t^2 - 1.23t.$$

The rate of change of position is usually called *speed* or *velocity*. Find the speed of the rocket at time $t = 123$ seconds after launch and also the height of the rocket at that time. Now use the differential to estimate the height at 123.1 seconds, 124 seconds, and 130 seconds. Calculate the correct values for the height at those times and exhibit them in a table with the error calculated for each estimate. Sketch a graph of $h(t)$.

*5. The speed $s(t)$ of a race car for the first 30 seconds after the start of the race is given in miles per hour after t seconds by $s(t) = 29.61\sqrt{t} - 0.173t$. Use a differential to estimate the *distance* (not the *speed*) traveled during the 10th second, the 20th second,

46

the 30th second (give your answers in feet; one mile is 5280 feet).
Sketch a graph of $s(t)$.

6. A certain sort of bacteria is known to multiply under lab conditions so that the area of its colony at t days after inoculation
is given in square centimeters by $A(t) = 1 + t + \frac{t^2}{2!} + \frac{t^3}{3!} + \frac{t^4}{4!} + \frac{t^5}{5!}$.
Sketch a graph of the growth during the first week after an inocula-
tion of one cm^2 with a population density of 10,000 individuals per
cm^2. Use differentials to approximate the number of individuals born
during the first hour of the fifth day, during the first minute of
the fifth day, during the first hour of the seventh day, and during
the first minute of the seventh day (remember to convert all time
intervals to decimal days). Calculate the exact population growth
during the first hour and the first minute of the seventh day and
compute the error in these two approximations by differential. (It
speeds computations to notice that after calculating $\frac{t^4}{4!}$, for example,
you only need to multiply by $\frac{t}{5}$ to obtain $\frac{t^5}{5!}$.)

*7. Graph on the same sheet of paper the functions x^3, $x^3 + 1$, $(x+1)^3$,
$x^3 + x$. Show graphically how to find, for each of these functions,
the value of the inverse function for the argument $x = 4$. Estimate
these values of inverse functions from your graph to 2 decimal places.
Now formulate a method to calculate the value of each of these inverse
functions at 4 and do so on your machine. Is there a conceptual difference in the four methods? A practical difference?

8. A farmer paces off one edge of a square field to measure it roughly. If he measures 116 yards and believes he is accurate within 2%, what size error is possible in his calculation of the area of the field (use a differential)? What percent error is that in the area?

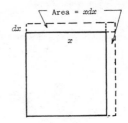

Figure 4.8

*9. A tuna can is to be made of steel .007 inches thick. It is to be 3 inches in diameter and 3 inches high. Calculate the number of cubic centimeters (cc) of steel that will be required (1 in = 2.54 cm). (Ignore the rims of the can.) Do this by use of a differential; then calculate the weight of steel required if that metal has a specific gravity of 7.78 (that is, the steel weighs 7.78 g per cc). Finally, find the weight of the steel in ounces (1 oz = 28.349523 g).

10. Let $f(x) = x^2 + 3$ and $g(x) = \sqrt{x} - 1/x$; calculate the difference quotients of $f(x)g(x)$ at $x = 2$ for increments of x of 0.1, 0.01, 0.001, 0.0001, ..., to find their limit, which is the derivative $(fg)'(2)$. Then evaluate $f(2)g'(2) + f'(2)g(2)$ as a check.

11. Follow the instructions of Exercise 10 to evaluate the limit of the difference quotients of the quotient function $\frac{f}{g}(x)$.

12. The concept of the derivative as a rate of change is used in business economics under the names *marginal cost* and *marginal profit*. As an example, suppose a small distilling unit in an oil refinery has fuel costs associated with its operation as follows: $20 to start it up plus $0.0027 per gallon distilled. Also, experience has shown that labor costs to run x gallons through the still are roughly

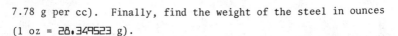

$\sqrt{x}/10$ dollars. The raw material to make one gallon of thinner costs $0.13. Thus the overall cost in dollars of filling an order for x gallons of paint thinner are

$$C(x) = 20 + 0.1327x + \sqrt{x}/10.$$

An order for 20,000 gallons would cost $2688.14. The added cost of producing one more gallon of thinner at the same time is the *marginal cost per gallon for* 20,000 *gallon orders*. This is

$$C'(20,000) = 0.1327 + 1/20\sqrt{20000}$$

Use a differential to estimate the added cost of adding 125 gallons to a still run of 20,000 gallons. Then compute the added cost as $C(20125) - C(20000)$ and compare the two figures. What error arose from the use of the marginal cost in a differential to estimate the added cost?

Next, estimate the added cost of distilling an added 125 gallons with a 1100 gallon order and compare your estimate with the computed cost as before.

13. A certain drug is found to raise human body temperature according to the formula $T(D) = 1.81 \, D^2 - D^3/3$. Here D is the dosage in grams in the range $0 \leq D \leq 3.5$, and $T(D)$ is the Fahrenheit change in body temperature due to that dosage (when there is no trace of the drug in the body to begin).

Find the dosage at which the body has maximal sensitivity to this drug. That is, find the dosage at which the greatest change in temperature results from a small change, say 10 mg, in the dose.

Problems

P1. Give a proof that $\lim_{dx \to 0} \dfrac{dy - \Delta y}{dx} = 0$, where $\Delta y = f(x+dx) - f(x)$ is the real increment in values of a function f and dy is the differential estimate of that increment, for a change dx in the argument. This theorem says that, in the limit as $dx \to 0$, the error in the estimate dy of Δy goes to 0 faster than dx does. Illustrate your theorem for the function $f(x) = x^2 + 2x + 3$ at the point 4 by calculating the values of $\frac{dy - \Delta y}{dx}$ at successively smaller values 0.1, 0.01, 0.001, 0.0001, ..., for dx.

*P2. What is the error in taking the ancient approximation $\pi \doteq \frac{22}{7}$? Now suppose you calculate π^3 using this approximation: use differentials to estimate the error caused by use of $\frac{22}{7}$ for π. Then do a similar job on the function $\pi^3 - 3\pi^2$. To how many decimal places must π be known to compute $\pi^3 - 3\pi^2$ accurately to the fifth decimal place (let this mean an error $< 10^{-5}/2 = 5 \times 10^{-6}$)? To how many decimal places must π be known to compute $\sqrt[3]{\pi}$ accurately to the fifth place?

*P3. A ball is thrown upward at 37.68 ft/sec from a 42.1 ft rooftop so that its height in feet above ground level after t seconds is

$$h(t) = 42.1 + 37.68t - 16t^2.$$

Find the time t when the ball reaches its maximal height (that's when it stops for an instant, so its speed is 0). What is that height? When does the ball arrive back at roof height, and how fast is it going then? How far does it fall in the 0.1 second after it passes roof height? When does it reach the ground? How fast is it going then? How high was it 0.1 second earlier? Illustrate all this by a sketched graph.

P4. Sketch a graph of the function $f(x) = \sqrt{x+1} - 1$ for positive arguments x. Imagine now that a line from the point (0, 1.27) on the y-axis just touches the graph of f at a single point. Find that point.

*P5. The *second derivative* $f''(a)$ of a function f at a point a is the derivative at a of the derivative function f' of f. It is defined by a limit

$$f''(a) = \lim_{h \to 0} \frac{f'(a+h) - f'(a)}{h}.$$

The numbers $f'(a+h)$ and $f'(a)$ are defined as limits, too:

$$f'(a+h) = \lim_{k\to 0} \frac{f(a+h+k) - f(a+h)}{k}$$

$$f'(a) = \lim_{k\to 0} \frac{f(a+k) - f(a)}{k}$$

Since both $h \to 0$ and $k \to 0$ in these limits, we may attempt to calculate $f''(a)$ by setting $h = k$ to get

$$f''(a) = \lim_{h\to 0} \frac{f(a+2h) - 2f(a+h) + f(a)}{h^2} \ .$$

Test this recipe on the function $f(x) = x^2 + 2x - 3$ for which we know $f''(a) = 2$, regardless of the value of a. Try values of a = 0, 1, 7, and let h successively take on values 10^{-i}.

Next, attempt to compute the second derivative of the function $f(x) = 67.89^x$ (see Problem P5, Ch. 3) at $a = 0$ by means of the double limit above (use $h = 1/2, 1/4, 1/8, \ldots$).

P6. The *Leaf of Descartes* is the set of points (x,y) in the plane for which $x^3 + y^3 - 3axy = 0$, where a is a scaling parameter, which we shall take to be $a = 1$.

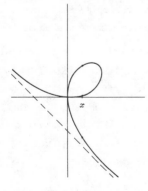

Figure 4.9

Suppose we wish to understand this curve at a point corresponding to, say, $x = 3/4$. We substitute 3/4 for x in the equation of the curve and solve the resulting cubic equation for y using algorithms from Chapter 2. It is clear from the graph that there will be three solutions $y_1(3/4)$, $y_2(3/4)$, and $y_3(3/4)$ to this cubic equation. Let us choose the middle one y_2, $0 < y_2 < 3/4$. The problem we now set is to find the slope of the tangent line to the graph at this point $[3/4, y_2(3/4)]$.

51

Find this slope by calculating the values of the function $y_2(x)$ at other points near to $[3/4, y_2(3/4)]$ on the graph, points $[3/4+h, y_2(3/4+h)]$ for values of $h = 10^{-3}$, 10^{-4}, 10^{-5}. Then compute the difference quotients. Can you think of another method of finding this slope?

*P7. Problem P5 in Chapter 3 asked for an evaluation of the limit

$$\lim_{x \to 0} \frac{67.89^x - 1}{x},$$

which we now know to be the derivative at 0 of the function 67.89^x. A hint suggested that x_i be taken to be $1/2^i$, and convergence was obtained only in the first 2 or 3 digits. With our new geometric picture of derivatives we can suppose that the chord from $(-x, 67.89^{-x})$ to $(x, 67.89^x)$ would have slope more nearly equal to the tangent at 67.89^0 than either of the shorter chords from the center. Make a table to evaluate this limit by evaluating

$$\lim_{x \to 0} \frac{67.89^x - 67.89^{-x}}{2x}.$$

Did you get greater accuracy? Make a sketch to display this technique. Give an arithmetic agrument that the two limits are equal.

P8. *Newton's method* for finding a zero of $f(x)$ uses the differential approximation df at $f(x_i)$ to find x_{i+1}: the derivative $f'(x_i)$ is the slope of the tangent line $y = (x-x_i)f'(x_i) + f(x_i)$, which is an approximation to the graph of f at $(x_i)f(x_i)$. The tangent line intersects the x-axis when $x = x_i - f(x_i)/f'(x_i)$; we take this value of x as the next estimate x_{i+1} for the zero of f. A starting guess x_0 must always be made.

Discuss this new method for finding zeros in case the function $f(x) = x^2 - a$ for some positive number a. Be sure to compare it to the techniques of Chapter 1. Next, use Newton's method to solve the

equation $x^3 - 3x - 1 = 0$, which we
examined in Chapter 2. Make a table of
your results corresponding to Table 2.1
and another like Table 2.3 (find only z_2
and z_3). Is this method better in the
sense that it requires fewer iterations?
Does it require fewer arithmetic operations, so that it is a faster technique
for you on your machine?

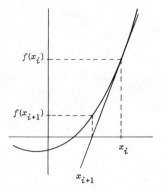

Figure 4.10

Over what interval of starting values x_0 would Newton's method
converge to z_1? Try starting with x_0 very near one end of this
interval; discuss the speed of convergence in this case.

P9. Describe at least one plausible situation in a field of your
own current interest where differentials and the notion of derivative
may be applied to get a useful numerical solution. Read about such
real-life situations by surveying a current issue of an appropriate
journal in your field. (See the Bibliography for some suggested
journal titles.)

Answers to Starred Exercises and Problems

Exercises   1a.   0.
             1b.   0.0235702
             1c.   -5.
             1d.   0.0312500
              5.   135 ft   during 10th second
                   189 ft   during 20th second
                   230 ft   during 30th second
              7.   The value at $x = 4$ for the inverse of
                   x^3 is 1.5874011, of x^3+1 is 1.4422496,
                   of $(x+1)^3$ is 0.5874011, and of x^3+x is
                   1.3787967.
              9.   1.34 oz

Problems   P2.   6-place accuracy for π assures 5-place accuracy for $\pi^3 - 3\pi^2$.

3-place accuracy for π assures 5-place accuracy for $\sqrt[3]{\pi}$

P3.   maximal height is $h(1.1775) = 62.2841$ ft

P5.   17.790585

P7.   4.2178887 = ln 67.89

5

MAXIMA, MINIMA, AND THE MEAN VALUE THEOREM

INTRODUCTION

Many everyday problems in the biological, social, and physical
sciences require that we find the exact situation in which some quantity is maximal or minimal. For instance, a stamping machine should
make coins rapidly; but when it runs too fast, the rate at which
inspectors reject faulty stampings increases and profits are dimin-
ished. Thus there is an optimal operating speed at which profit is
maximal.

Another example is a chemical reaction, which proceeds ever
more rapidly as temperature is raised, saving on equipment time and
labor. However, unwanted by-products may increase with higher tempera-
tures, and this may downgrade the value of the product. Again, there
will be an optimal temperature for most efficient operation.

We now study some simple Examples in which the calculus can
show us where the maximal and minimal values lie. We shall also
examine the use of the Mean Value Theorem to find maximal and minimal
limits on the amount of change in the values of a function. Many

Exercises demonstrate applications of these ideas. The first Problem is another application but a more difficult one. Another Problem discusses the limitations of the use of differentials. Also, the convergence of iterative functions in algorithmic methods is established by means of the Mean Value Theorem in a Problem.

EXAMPLE: A MINIMAL FENCE

Suppose a rancher wants to fence off a small rectangular field of, say, $2\frac{1}{2}$ acres along the inside of a long fence that already exists. Two and a half acres is $2.5 \times 5280^2/640 = 108,900$ square feet, or 12,100 square yards. The rancher's first thought is to use three equal sides of fencing, each of $\sqrt{12,100} = 110$ yd length, making a square fenced region (Figure 5.1).

Figure 5.1

This seems to him better than to fence off a long narrow rectangle in either of the two possible extreme ways illustrated in Figure 5.2, since it will clearly require less new fencing to make the new area square. True enough, but is the square the best he can do? To determine this, let x stand for the length of the side of the rectangle that touches the existing fence, so that the total length of fencing the rancher will require is

$$f(x) = 2x + \frac{12100}{x} \text{ yards.}$$

Obviously the values of f become enormous as $x \to 0$ or as x itself gets enormous (x must be a positive number).

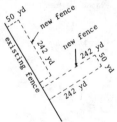

Figure 5.2

Thus there is some minimal value $f(x_0)$ where the amount of fencing required is least (see Figure 5.3). This value x_0 of the argument gives the correct configuration of the rectangle. But how are we to find x_0? Well, it is clear from the graph of this function that the slope of the tangent line at $f(x)$ is negative if $x < x_0$ and is positive if $x > x_0$. If we solve the equation $f'(x) = 2 - 12100/x^2 = 0$ or

$$2x^2 - 12100 = 0$$
$$x^2 - 6050 = 0$$
$$x^2 = 6050$$
$$x = \sqrt{6050} = 77.781745,$$

we see that only at $x_0 = 77.79$ or about 77 yd 28 in is the tangent line horizontal (for positive x). This value of x_0 is then the

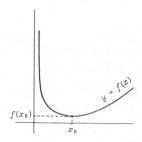

Figure 5.3

unique length for the side touching the existing fence to minimize the total length $f(x_0) = 311.12698$ yd, about 311 yd 5 in. This is, of course, a considerable improvement on the 330 yd requirement for the square region; the rancher can save 6% of his fencing cost this way.

The THEOREM goes as follows: *if* f *is a continuous function on a closed interval* [a,b], *then* f *actually attains a maximum value at some point* x_0 *in this interval* (x_0 *may not be unique*). *This point* x_0 *may be one of the endpoints,* a *or* b, *or it may be a point where* f *is not differentiable; otherwise,* f'(x_0) = 0. The same holds for minimum values of f: there always is at least one point x_1 such that $f(x_1)$ is minimal on [a,b] and

    (i) x_1 is an end point of the interval, or

    (ii) f is not differentiable at x_1, or

  (iii) $f'(x_1) = 0$.

Hence, to find all the points where f is maximal or minimal, consider these three classes of points. Of course, f does not necessarily take on extreme values at all of these points; you must evaluate the

57

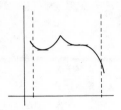

Figure 5.4

function at all the points of these three classes to find its extremes. Figure 5.4 depicts points of each type.

THE MEAN VALUE THEOREM

In Chapter 4 we saw that the use of differentials in the estimation of errors offers a method for simple yet surprisingly accurate approximations of the changes in values of a function that correspond to small errors in its argument. There are occasions, though, where this method is not satisfactory. For example, we may be unwilling to accept the approximation by a differential because we are unsure it is, in the particular case at hand, very accurate at all. Or we may be more interested in knowing *with certainty* an upper bound, some maximal size limit on the error.

In these cases, a related method works. The MEAN VALUE THEOREM asserts that *if a function* f *is differentiable at each point of a closed interval* $[a,b]$, *then there is some point* c, a < c < b, *with* $\frac{f(b) - f(a)}{b - a}$ = f'(c). That is, there is a point c inside the interval where the slope of the tangent line to the graph of f is equal to the slope of the chord of the graph over the whole interval (see Figure 5.5).

Figure 5.5

EXAMPLE: CAR SPEED

A physical example will illustrate this. If $f(x)$ is the distance traveled by time x, then the difference quotient is the *average speed*

58

from time a to time b and $f'(c)$ is the instantaneous speed. The theorem then asserts that if during a race a Lotus car averaged 243 kilometers per hour, then at some moment of the race the Lotus was traveling at exactly 243 kilometers per hour - not very surprising, is it?

If we rewrite the conclusion of the Mean Value Theorem (the MVT)

$$f(b) - f(a) = (b-a)f'(c),$$

it says that the exact change in value of f is given by the value of the differential at some point c between a and b. In our use of differentials we merely used the differential at one of the endpoints, a or b, to approximate this change in functional values. In a sense, the MVT is little help, since it does not tell us how to find c.

EXAMPLE: PAINTING A CUBE

Let us reexamine our Example of Chapter 4. There we wished to know the volume of a paint film 0.012 inches thick on a metal cube with edges of length 3.456 inches. The function we work with is $V(x) = x^3$, the volume of a cube of edge length x. The volume of the paint film is $(3.456+0.024)^3 - 3.456^3$; the MVT says this difference in values of V is $0.024V'(c)$ for some c with $3.456 < c < 3.456 + 0.024$ and $V'(c) = 3c^2$. Hence we may take the minimum $V'(3.456) = 35.831808$ and the maximum $V'(3.456+0.024) = 36.3312$ of V' on this interval to see that $35.831808 \le V'(c) \le 36.3312$. Hence $0.8599634 \le 0.024V'(c) \le 0.8719488$ cubic inches, and we now have more than an estimate for the volume of paint film: we have an interval of values within which the correct answer *must* lie. We now not only have the estimate, but also we know how close it is. (The correct answer is 0.8659492.)

EXERCISES

1. Find the maximum and minimum values for each function:

  *a. $x^3 - 5x$ on $[-2,2]$ *c. $2x^3 + 3x^2 - 6x - 7$ on $[-2,1]$

  *b. $x^3 + x^2 + 1$ on $[-2,1]$ *d. $x^3 - 12x^2 + 42x$ on $[1,6]$

2. Consider a maximal problem associated with the rancher's fence.

Let the rancher have exactly 311.12698 yards of fencing (left over, perhaps, from another project) with which he wishes to enclose the largest possible rectangular region, using the long existing fence as one edge. If x represents the length of each of the two edges that touch the existing fence, what is the area $A(x)$ of the rectangle as a function of x? Clearly we must have $0 \leq x \leq 311.12698/2$. What is the value of $A(x)$ at the extreme possible values of x (endpoints)? Is $A(x)$ differentiable everywhere? For which x does $A(x)$ take on extreme values? Which of these values are maxima?

Compare your results with the conclusions of our example and discuss the similarities.

*3. Suppose a fisherman is in a boat 320 meters out from the river's edge, and his house is 1100 meters down the river from the closest point on the shore to the boat. If he observes that his house is afire and wishes to get home as fast as he can, what path should he follow? Assume that he can row his boat at a rate of 1.1 meters per

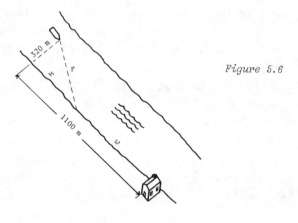

Figure 5.6

second and run along the river bank at 5.3 meters per second. We have sketched the situation depicting his landfall x meters down the river toward his house. Find the distances r and w as functions of x; then express the time taken for each leg of his trip and then total trip time $T(x)$. Now find the minimal value of total time by

60

considering all the possible values of x for which your function T might have extreme values.

*4. A cylindrical catfood can is to be designed to use the minimal amount of sheet metal for its volume, which is to be 300 cc. Express the top and bottom areas as a function of the radius of the cylinder,

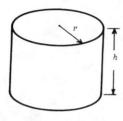

then express the height also as a function of the radius, remembering that the volume is fixed. Now find the area $A(r)$ as a function of the radius r and minimize. What is the appropriate radius and the minimal area?

Figure 5.7

5. Suppose you wish to make a 300 cc cyclindrical metal open-top cup of minimal area (see Exercise 4). What shape should it be?

*6. Suppose, in Exercise 4, the catfood can is to be of maximal volume for a fixed area of 248.08207 cm². Find the appropriate radius for maximal volume, and calculate that volume.

7. Find the area of the largest rectangle that can be inscribed in the ellipse $x^2 + 2y^2 = 3$. (You may assume that the sides of the rectangle are parallel to the coordinate axes.)

*8. Calculate by means of the MVT the maximal error in $f(x) = x^3 - 9x - 2$ if we use 22/7 to calculate $f(\pi)$. That is, find the maximal value of f' on the interval $[\pi, 22/7]$ and use this with the MVT to calculate an upper bound for $f(22/7) - f(\pi)$. Then express this error as a percentage of $f(\pi)$. Next, calculate the precise error; again report this error as a number and also as a percentage of the correct value $f(\pi)$. What is the error and percent error in the approximation 22/7 for π? (Compare Problem 2, Ch. 4.)

9. Use the MVT to calculate upper and lower limits on the volume of metal required to make a cylindrical can of radius 2.87 inches and height 6.53 inches if the sheet metal used is 0.00814 inches thick.

PROBLEMS

*P1. Find the point on the parabola $y = x^2$ that is nearest to the point (3,2).

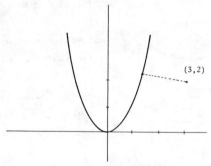

Figure 5.8

P2. Use the MVT to give upper and lower bounds on an error in the calculated volume of a cube of metal that is due to an erroneous

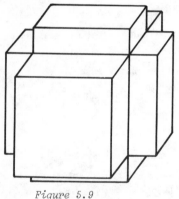

Figure 5.9

recording of the measurement of an edge as 2.3 cm when the correct measurement is 2.0 cm. Also give the error in the calculated volume of the cube predicted by use of differentials for an error of 0.3 in a measurement of 2.0. What is the real error? Discuss the failure of the differential to predict the error even approximately in this case.

P3. Let an algorithm be given by $x_{n+1} = \varphi(x_n)$, and assume that $\varphi(y) = y$. Define the nth error ε_n for the successive approximations $x_0, x_1, \ldots x_n, \ldots$ for y to be $\varepsilon_n = y - x_n$. Suppose $a < y < b$ and that $M < 1$ is a bound for $|\varphi'(x)|$ on $[a,b]$, so that for every x between a and b we have $1 > M \geq |\varphi'(x)|$. Use the MVT to show that $|\varepsilon_n| \leq |M\varepsilon_{n-1}|$ and then from that fact prove that the sequence of approximations $x_0, x_1, \ldots$ does indeed converge to y.

62

Answers to Starred Exercises and Problems

Exercises   1a.   ±4.3033148

1b.   ±3

1c.   -9.0901699 and 2.0901699

1d.   31 and 45.656854

3. x = 67.89 m

4. r = 3.6278317 cm

6. r = 3.6278317 cm

8.   The error for $f(x)$ is no greater than
0.0260898 or 3.6%.

Problems   P1. x = 1.5674684

6

TRIGONOMETRIC FUNCTIONS

INTRODUCTION

More than 4000 years ago the Egyptians used ropes knotted in lengths
with the ratios 3 to 4 to 5 to form the sides of a triangle in order
to determine a right angle. Buildings, and probably the pyramids,
were thus erected with the use of elementary, practical trigonometry.
Carpenters use this same trick today.

By the time of Christ, the uses of trig had expanded from building and surveying to astronomy. Hipparchus of Rhodes made up trig
tables then and did spherical trigonometry. Today tide tables, moon
shots, and television sets depend on our knowledge of trig.

In this chapter we will review angles and the definitions of
the trig functions, together with their uses in similar triangles.
We will also determine and study the derivatives of the trig functions.
These derivatives are next applied via the chain rule to establish
the derivatives of the inverse trig functions. Examples and Exercises
numerically evaluate these derivatives at particular values of their
arguments. There are also iterative root-finding algorithms involving

trig functions. These allow us to see the limit processes in operation with these functions, which are indeed our old friends among
transcendental functions.

We will explore these topics further in the Problems, as well as
spherical trig, continued fractions, and some modifications of difference quotients that offer numerical advantages on our calculators.

ANGLES

An angle is a geometric sort of thing, but the measure of an angle
is a number. If X stands for the geometric angle $\diamondsuit P O Q$ in
Figure 6.1, then the measure x
of X expresses the ratio that
X bears to a whole circle, the
proportion that the slice $\diamondsuit P O Q$
is of a whole pie. If the
measure of the whole circle is

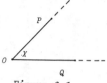

Figure 6.1

taken to be 360, then a third of a circle, for instance, has measure
120 and a quarter circle or right angle has measure 90. This method
of measuring angles is called *degree measure*.

A protractor can be used to measure angles. A half disc of
metal has its circular edge divided into 180 equal pieces so each
piece is 1/360 of the circumference of the whole circle (see Figure
6.2). This is compared to an angle $\diamondsuit P O Q$, and the degree measure

Figure 6.2

is read off. If the protractor's radius $\overline{OP}$ is used as a unit of
measurement of length (so $r = 1$), then the length of the semicircular
arc is π units, and each division on it has length $\pi/180 = 0.0174533$.
Another scale for measuring angles is given when the arc of the
protractor is marked off directly in lengths, using the units
for measurement of length in which $r = 1$. An angle is then

measured in the same way as before by the protractor (see Figure 6.3).
The resulting number is called the *radian measure* of the angle. The
angle whose radian measure is 1 has degree measure 180/π = 57.295780.
This fact is often expressed by saying "one radian equals 57.295780
degrees." We shall use only
radian measure in this book
unless we specify otherwise in
a particular case.

Figure 6.3

TRIG FUNCTIONS

The trigonometric functions are functions of numbers, and they have
numbers as values. Their definitions, though, are geometric in flavor.
The number *sin* x is defined as the length QR of the segment of $\overline{QR}$ on a
protractor whose radius is 1, if the arc from P to Q is of length x,

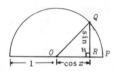

Figure 6.4

$\overparen{PQ}$ = x (see Figure 6.4). This
last statement, $\overparen{PQ}$ = x, is the
same as the requirement that
$\sphericalangle P \, O \, Q$ have radian measure x.
The cosine function is defined
by setting cos x = OR, the
length of the segment $\overline{OR}$. The other trig functions are then defined
algebraically from the sin and cos functions:

$$\tan x = \frac{\sin x}{\cos x} \, , \quad \cot x = \frac{\cos x}{\sin x} \, , \quad \sec x = \frac{1}{\cos x} \, , \quad \csc x = \frac{1}{\sin x}$$

If the protractor had
been made to be a complete
disc, instead of just half
of one, then sin x could
be defined for any number
x between 0 and 2π by
going counterclockwise
around the

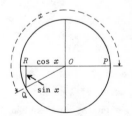

Figure 6.5

protractor from P a distance x to locate Q (see Figure 6.5). The number $\sin x$ is still the distance QR, but it is to have a negative signature[†] whenever Q lies below R instead of above (that is, whenever $\pi < x < 2\pi$). Similarly, $\cos x$ is to be the distance OR, but it is to have a negative signature whenever R is to the left of 0 (that is, whenever $\pi/2 < x < 3\pi/2$). Pythagoras' Theorem says that $(OQ)^2 = (OR)^2 + (RQ)^2$. Since $OQ = 1$ on our protractor, this yields

$$\sin^2 x + \cos^2 x = 1.$$

TRIANGLES

Two triangles are similar if they have all their angles the same; the fundamental rule about similar triangles is $a/d = b/e = c/f$, where

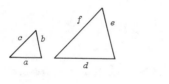

Figure 6.6   Figure 6.7

each of these letters from Figure 6.6 stands for the length of the side it is near. This is used as follows: suppose the (radian) measure of the angle at B in Figure 6.7 is x, and we let the angle at C be a right angle. The angle at A must then have (radian) measure $\pi/2 - x$. Therefore the triangle in Figure 6.7 is similar to the triangle in Figure 6.4, which has its hypotenuse of length 1. It follows that the sides of ABC are related by $b/c = \sin x$, $a/c = \cos x$ and $b/a = \tan x$. If Figure 6.7 represented a tree, for example, then a could be measured and b calculated as $b = a \tan x$.

[†]The signature of a number is its sign, plus or minus. We use "signature" here since "sign" sounds like "sin."

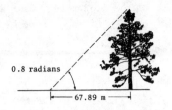

Figure 6.8

In Figure 6.8 the height of the tree is ᒷᑍ.ᑍᑌᔕᔕᔕᒷ = 67.89 tan (0.8) meters.

EXAMPLE: THE DERIVATIVE FOR sin x

The derivative of the function sin x at $x = 0$ is

$$\lim_{x \to 0} \frac{\sin x - \sin 0}{x} = \lim_{x \to 0} \frac{\sin x}{x}.$$

It is intuitively clear from the definition (see Figure 6.4) that this limit is 1. That is, in the limit, as Q gets close to P, the length of the arc x and the length of the perpendicular QR become equal. This is readily verified on a machine that computes the sin function. We list the results in Table 6.1 (remember sin $(-x)$ = -sin x, so $\frac{\sin(-x)}{-x} = \frac{\sin x}{x}$.

TABLE 6.1

x	sin x	$\dfrac{\sin x}{x}$
0.1	0.0998334	0.9983342
0.01	0.0099998	0.9999833
0.001		
0.0001	0.0001	1.

The derivative of the function sin x is

$$\frac{d}{dx} \sin x = \cos x.$$

68

This may be easily shown by direct evaluation of the limit, using trig identities and the limit that is calculated above. We content our-selves with a direct calculation of the derivative of sin x at $x =$ 0.2345 radians; this means we must evaluate the limit of the differ-ence quotient

$$\lim_{h \to 0} \frac{\sin (h + 0.2345) - \sin (0.2345)}{h} .$$

The results appear in Table 6.2.

TABLE 6.2

h	Difference Quotient
0.1	0.9594022
0.01	0.9714526
0.001	
0.0001	
0.00001	0.9725
0.000001	0.972
	.
	.
	.
−0.0001	0.97265
−0.001	

The correct result is cos (0.2345) = 0.9726306, which we have to 4 decimal places for $h = -10^{-4}$. The anomalous result for $h = 10^{-5}$ is due to digits that our machine dropped in calculation, so this method will not yield more than 4-place accuracy on a 10-digit machine. On an 8-digit machine there is 3 place accuracy for $h = 10^{-3}$. See Exer-cise 12 and Problem 6 for better numerical methods.

DERIVATIVES FOR TRIG FUNCTIONS

The derivative of the function cos x is

69

$$\frac{d}{dx} \cos x = -\sin x.$$

We relegate to the Exercises the calculation of this derivative at certain points. The other trig functions are algebraically defined from the functions sin x and cos x. Hence their derivatives may be taken by means of rules for differentiation:

$$\frac{d}{dx} \tan x = \sec^2 x,$$

$$\frac{d}{dx} \cot x = -\csc^2 x,$$

$$\frac{d}{dx} \sec x = \sec x \tan x,$$

$$\frac{d}{dx} \csc x = -\csc x \cot x.$$

EXAMPLE: $f(x) = x \sin x - 1$

Consider the function $f(x) = x \sin x - 1$. Since sin $(-x) = -\sin x$, $f(-x) = f(x)$. Also we know that $f(0) = f(\pi) = -1$, and sin $(\pi/2) = 1$ so $f(\pi/2) = \pi/2 - 1 > 0$. Thus there is a zero of f between 0 and $\pi/2$ and another zero between $\pi/2$ and π (see Fig. 6.9). We solve for a zero by observing

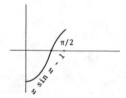

Figure 6.9

$$f(x) = x \sin x - 1 = 0$$
$$x \sin x = 1$$
$$x = 1/\sin x.$$

70

Thus we try the algorithm $x_{i+1} = 1/\sin x_i$, starting with $x_0 = 1$. Since this process converges slowly, we dispense with our usual table of functional values. Instead we rapidly and repeatedly compute from x_i to $\sin x_i$ to $x_{i+1} = 1/\sin x_i$: our limit is $x_{29} = 1.1141571$, and $f(x_{29}) = -0.0000001$.

INVERSE TRIG FUNCTIONS

The function *arcsin* x is an inverse function for the sin x function. That is, for each number x between -1 and 1, the number y = arcsin x is a solution y of the problem sin y = x. There are, of course, many

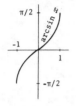

Figure 6.10

other solutions, but the custom is to choose y between $-\pi/2$ and $\pi/2$ (see Figure 6.10). The other values of y may be found from the identities sin $(-y)$ = -sin y, sin $(\pi+y)$ = -sin y, and of course sin $(2\pi+y)$ = sin y. For instance, sin $(\pi-y)$ = sin y. Hence the function $f(x) = \pi$ - arcsin x is a different function than arcsin x, but it is another inverse function for sin x, so sin (arcsin x) = x and also sin $[f(x)]$ = x.

The values for the function arctan x are again usually taken to be in $[-\pi/2, \pi/2]$, whereas the function arccos x usually has range $[0,\pi]$. Other values for these inverse functions can be understood by inspection of the graphs of tan y and cos y, respectively, or by use of the identities cos $(\pi+y)$ = -cos y, cos $(-y)$ = cos y, and tan $(\pi+y)$ = tan y, tan $(-y)$ = -tan y.

The derivatives of the inverse trig functions are easily taken by the chain rule. As an example, sin (arcsin x) = x so

$$\frac{d}{dx} \sin (\text{arcsin } x) = 1 = \cos (\text{arcsin } x) \frac{d}{dx} \text{arcsin } x.$$

But

$$\cos (\text{arcsin } x) = \sqrt{1-\sin^2(\text{arcsin } x)} = \sqrt{1-x^2};$$

this means

71

$$\frac{d}{dx} \arcsin x = \frac{1}{\sqrt{1-x^2}} .$$

The other derivatives are:

$$\frac{d}{dx} \arccos x = \frac{-1}{\sqrt{1-x^2}}$$

$$\frac{d}{dx} \arctan x = \frac{1}{1+x^2}$$

$$\frac{d}{dx} \arccot x = \frac{-1}{1+x^2}$$

$$\frac{d}{dx} \arcsec x = \frac{1}{x\sqrt{x^2-1}}$$

$$\frac{d}{dx} \arccsc x = \frac{-1}{x\sqrt{x^2-1}}$$

EXAMPLE: $f(x) = 2 \arcsin x - 3x$

Suppose we wish to find a minimal value for $f(x) = 2 \arcsin x - 3x$.
This function is zero at $x = 0$, and $f(1) = \pi - 3$ is positive. The
derivative $f'(x) = 2(1-x^2)^{-\frac{1}{2}} - 3$ and $f'(0) = -1$, so f is decreasing at
$x = 0$. Hence f has a minimal value between 0 and 1. To find that
minimum, set $f'(x)$ equal to 0 and solve:

$$2(1-x^2)^{-\frac{1}{2}} = 3$$

$$1-x^2 = 4/9$$

$$x^2 = 5/9$$

$$x = \sqrt{5/9} = 0.7453560.$$

The minimal value for the function at this point is $f(\sqrt{5/9}) =$
-0.5539306. For fun you may wish to find the zero of f that is near
$x = 1$.

72

EXERCISES

1. Consider the right triangle with sides of length 3, 4, and 5, and let α denote its smallest angle, which is opposite the smallest side, so α = 0.6435011. For each trig function indicated below, first give its value at α. Next, estimate the value of the derivative of the given function at α by computing its difference quotient at $h = 10^{-5}$. Finally, compare your estimated value for the derivative at α with the correct 7-place value that you calculate for the theoretical derivative function by inspection of the triangle.

*a. sin   *c. tan   *e. sec

 b. cos   d. cot   f. csc

2. Evaluate the derivative of cos x at 0: find $\lim\limits_{h\to 0} \dfrac{\cos(h)-1}{h}$ by making a table of values at $h = \pm 0.1, \pm 0.01, \ldots$.

3. Evaluate the derivative of cos x at 0.54321. Make a table of values of the difference quotient

$$\frac{\cos\ (h + 0.54321) - \cos\ (0.54321)}{h}$$

at $h = \pm 0.1, \pm 0.01, \ldots$. Discuss the numerical problem of the convergence to $-\sin(0.54321) = -0.5168866$.

4. Show that if the sin function were defined just as in Figure 6.4 except that its argument is agreed to be the degree measure of the angle, instead of radian measure, then $\lim\limits_{h\to 0} \dfrac{\sin h}{h} = \dfrac{\pi}{180}$. Do this first by direct evaluation of the limit, making a table of values of $\dfrac{\sin h}{h}$ for diminishing values of h. Then give a proof that this is as it should be.

*5. Make a table of values to evaluate $\lim\limits_{h\to 0} \dfrac{\sin 3h}{\sin 2h}$. Then give a proof that your limit is as it should be.

*6. Find the zero of $f(x) = x \sin x - 1$ between $\pi/2$ and π. First, test the algorithm $x_{i+1} = \dfrac{1}{\sin x_i}$, trying various values like 2.5, 2.6, 2.7, 2.8, 2.9, 3.0 for x_0, to see that it will not converge to this zero. Then try its inverse algorithm $x_{i+1} = \arcsin (1/x_i)$; this gives absurd values. Now draw a graph of the arcsin function and notice that the value we seek is not the one in $[-\pi/2,\ \pi/2]$ but rather the angle in $[\pi/2,\ 3\pi/2]$. Hence your algorithm should be $x_{i+1} = \pi - \arcsin 1/x_i$, starting with $x_0 = \pi/2$. Make a table of your results and check your answer.

7. Draw a graph of the function $f(x) = x \sin x - 1$ from our example, using the zero you have computed in Exercise 6 and the results of our Example. Evaluate the function from 0 to π in increments of 0.2 and label the x- and y-intercepts. In order to clarify the theory, lightly sketch in (perhaps in colors) the graphs of $g(x) = \sin x - 1$ and $h(x) = x - 1$, as well as the line $y = -1$. Can you figure out the shape of $f(x)$ from the other graphs?

*8. The *law of cosines* says that for any triangle with sides of lengths a, b, and c, and C the (measure of the) angle opposite side c,

$$c^2 = a^2 + b^2 - 2\ ab \cos C\ .$$

Given that $a = 67.89$, $b = 123.45$, and $C = 1.234$, find c. Next,

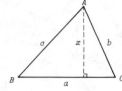

Figure 6.11

follow the diagram to compute $x = b \sin C$ and $\sin B = x/c$, so $B = \arcsin \dfrac{b \sin C}{c}$; then $A = \pi - B - C$. (The *law of sines* says

$$\frac{a}{\sin A} = \frac{b}{\sin B} = \frac{c}{\sin C}\ .)$$

Now use a differential to estimate how much error would result in c if the measurement of C had been off by 0.001.

*9. If all three sides of a triangle have known lengths, (the measure of) one angle may be computed from the law of cosines:

$$C = \arccos \frac{a^2 + b^2 - c^2}{2\ ab}$$

74

The other angles may then be found by use of the law of sines (Exercise 8). Do this for side lengths 1.23, 2.34, 3.45. Next, use a differential to estimate the error in B caused by an error of 0.0009 in transcribing C.

*10. An airplane is flying parallel to the coastline, 47.6 miles offshore and 521 miles per hour. There is an airport on shore at a bearing of $B = 34°56'$. How rapidly, in degrees per minute, is the bearing B changing? (Hints: B must be converted to radian measure

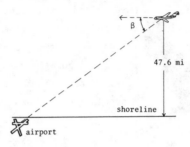

47.6 mi

shoreline

airport

Figure 6.12

before differentiation, as in Exercise 4. Then B and x are functions of time t, and $\frac{dx}{dt} = -521$ mph.)

*11. Sketch a graph of the function $f(x) = \cos x - 3x$: calculate and indicate on the graph the maximal and minimal values of f as well as its zero(s) on the interval $[0,1]$.

12. Show that $f'(x) = \lim\limits_{h\to 0} \frac{f(x+h)-f(x)}{h} = \lim\limits_{h\to 0} \frac{f(x)-f(x-h)}{h}$. Therefore the limit as $h\to 0$ of the average $\frac{1}{2}\left[\frac{f(x+h)-f(x)}{h} + \frac{f(x)-f(x-h)}{h}\right] = \frac{f(x+h)-f(x-h)}{2h}$ must also be $f'(x)$ (can you say why?). Repeat the calculation of the derivative of $\sin x$ at $x = 0.2345$, which is given in Table 6.2, but this time use the estimate $\frac{\sin(x+h)-\sin(x-h)}{2h}$ instead of the difference quotient $\frac{\sin(x+h)-\sin(x)}{h}$. Notice the greater accuracy. Sketch a graph that shows this new quotient as the slope of an appropriate chord of the graph of $\sin x$. On the same graph sketch the chords corresponding to increments of h and $-h$ in the difference

quotient. Do you see why the averaged quotient described above gives a better approximation for a given value of h than the difference quotient?

13. Many models for the study of population growth are formulated in terms of ordinary differential equations. In particular, the removal of members from a population at a constant rate can be modeled by such an equation containing the harvest rate as a parameter. Brauer and Sanchez (*Theoretical Population Biology*, in press) have solved a modification of the "logistic" equation of Lotka. Their study shows that for harvest rates E greater than a critical rate E_c, populations tend to zero in finite time. The extinction time T in years for a population is shown to be

$$T = 4(4aE-\lambda^2)^{-\frac{1}{2}} \arctan \left[\lambda(4aE-\lambda^2)^{-\frac{1}{2}}\right] .$$

Here the equilibrium population in the unharvested case is λ/a, and the critical rate of harvesting is $E_c = \lambda^2/4a$.

Observations suggest that with no harvesting the equilibrium population for sandhill cranes (*grus canadensis*) is about 194,600 and that the critical harvest is $E_c = 4,800$. Calculate the extinction times T for harvest rates E of 6000 and 12,770 per year.

Next, use a differential to estimate the change in extinction times due to an additional harvesting of 100, 200, 500, or 1000 more than 12,770 per year.

14. A distributor observes that his refrigerator *sales rate* cycles during the year to grow according to the formula

$$S(t) = 1.012^t (\sin\left[\frac{\pi}{12} (t + 1.7)\right]+6.2) .$$

Here t is measured in months and $S(t)$ is sales in hundreds of units per month. Use a differential to estimate *total sales* during the first week of July. (Be sure you use the derivative of the appropriate function. Consider the week to be 7/31 of the month.)

PROBLEMS

*P1. Find the maximum of $f(x) = x \sin x - 1$ on the interval $[0,\pi]$.
Do this by setting the derivative $f'(x) = 0$ and solving iteratively.
This is difficult; the algorithm $x_{i+1} = -\tan x_i$ does not converge,
so one must use the inverse algorithm $x_{i+1} = \arctan(-x_i)$. To have
this succeed, you will need to sketch a graph of the algorithm and
observe that the usual value of the arctan function is not the correct
one. Add your results to the graph made in Exercise 7.

P2. Graph the function $f(x) = x^2 - \sin x$ on the interval $[0,1]$.
Display the maxima, the minima, and the zeros that you calculate for
f.

*P3. A *spherical triangle* on the earth's surface is given by its
three vertices: the sides are measured by the angles they subtend
at the center of the earth, and the angles at the vertices are
measured between appropriate planes. The triangle itself may be
thought of as the three arcs of great circles connecting the vertices.
The surface distance from B to C along such an arc is 60 nautical

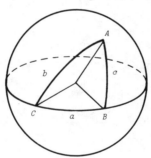

Figure 6.13

miles for each degree of the angle a. The law of cosines has two
forms here:

$$\cos a = \cos b \cos c + \sin b \sin c \cos A,$$

$$\cos A = -\cos B \cos C + \sin B \sin C \cos a,$$

and the law of sines becomes:

$$\frac{\sin a}{\sin A} = \frac{\sin b}{\sin B} = \frac{\sin c}{\sin C}.$$

Suppose two airplanes start at right angles to each other and fly until one has gone 300 nautical miles, the other 400. If the earth were flat, they would then be 500 nautical miles apart. How far apart are they on the earth? If the first plane has landed and the second is continuing at 400 nautical miles per hour, how fast does the radio bearing of the second plane seem to change to an observer in the first plane (in degrees per minute)?

P4. Suppose water pipe is to be carried (horizontally) down a hall 3.15 m wide and then into a hall 2.41 m wide that meets the first

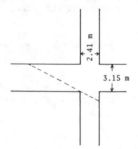

Figure 6.14

hall at right angles. What is the greatest length of pipe that can make the turn?

P5. A *continued fraction* is a set of directions for an algorithm. For example,

$$\cfrac{1}{2 + \cfrac{3}{4 + \cfrac{5}{6 + \cdot}}}$$

means the sequence of numbers 1/2, 1/(2 + 3/4), 1/[2 + 3/(4 + 5/6)], ... It may also be written as

$$\frac{1}{2+} \ \frac{3}{4+} \ \frac{5}{6+} \ \cdots \ .$$

These algorithms are, of course, nothing more than simple recipes using sums and products, quotients and differences. As such, they can express rational functions well. For instance, there is a continued fraction expansion of the function arctan x:

78

$$\arctan x = \frac{x}{1+} \frac{x^2}{3+} \frac{4x^2}{5+} \frac{9x^2}{7+} \frac{16x^2}{9+} \frac{25x^2}{11+} \cdots ,$$

which is valid for every number x. It is also written

$$\arctan x = \cfrac{x}{1 + \cfrac{x^2}{3 + \cfrac{4x^2}{5 + \cfrac{9x^2}{7 + \cfrac{16x^2}{9 + \cfrac{25x^2}{11 + \cdots}}}}}}$$

This expression refers to a sequence of functions:

$$f_1(x) = x,$$

$$f_2(x) = x/(1 + x^2/3) = 3x/(3 + x^2),$$

$$f_3(x) = (15x + 4x^3)/(15 + 9x^2), \ \ldots \ .$$

Evaluate these functions $f_1, f_2, \ldots, f_8$ at the points $x = 1$ and $x = 1/\sqrt{3}$. Tabulate your results and compare them with the limiting correct values $\arctan(1) = \pi/4$ and $\arctan(1/\sqrt{3}) = \pi/6$. (Hint: You can find the values of the functions by directly evaluating the continued fraction. You do not need to derive the expressions for the functions as quotients of polynomials.)

P6. The quotient $Q_2 = \dfrac{f(x+h)-f(x-h)}{2h}$ was shown in Exercise 12 to be a better way to estimate $f'(x)$ than is the difference quotient $Q_1 = \dfrac{f(x+h)-f(x)}{h}$ itself. While Q_2 was viewed in Exercise 12 as the slope of a chord from $f(x-h)$ to $f(x+h)$ on the graph of f, this new quotient may also be shown to be the derivative at x of the quadratic polynomial that fits the graph of f at the points $f(x-h)$, $f(x)$, and $f(x+h)$.

It is possible to fit a quartic polynomial to the graph of f at the five points x, $x\pm h$ and $x\pm 2h$. The derivative of this quartic polynomial can be shown to be

$$Q_3 = \frac{1}{12h} \left[f(x-2h) - 8f(x-h) + 8f(x+h) - f(x+2h) \right].$$

Construct a table similar to Table 6.2 using Q_3 in place of Q_1 to estimate the derivative of $\sin x$ at 0.2345. Compare your results with $\cos$ 0.2345. [For further discussion of this subject see the article by David A. Smith, Numerical differentiation for calculus students, *The American Mathematical Monthly*, 82 (1975), 284-87. (This is a standard form of reference to pages 284 through 287 of volume 82. These pages are in the March 1975 issue.)]

Answers to Starred Exercises and Problems

Exercises   1a. $\sin \alpha = 0.6$, difference quotient =
                  0.7999800, $\cos \alpha = 0.8 = 4/5$

            1c. $\tan \alpha = 0.75$, difference quotient =
                  1.5624800, $\sec^2 \alpha = 1.5625 = (5/4)^2$

            1e. $\sec \alpha = 1.25$, difference quotient =
                  0.9375 = $\sec \alpha \tan \alpha = 5/4 \times 3/4$

            5. $3/2$

            6.   2.7726047

            8. $c = 119.62310$, $B = 1.3423183$

            9. $C = 2.5937661$

            10.   7.0248650 deg/min

            11.   max at $f(0)$, zero at $f(0.3167508)$,
                  min at $f(1)$.

Problems   P1.   0.8197057

            P3.   499.59310 nautical miles apart

7

DEFINITE INTEGRALS

INTRODUCTION

We now begin a study of the second principal concept of the calculus: integration. Its origins go back to Arçhimedes, who thought of areas (and volumes) as being made up of tiny pieces, each of which was a triangle or square or other regular figure. This is just the way you would think about the area of a tiled patio with a curved boundary. Since you know the area of each square tile, you need only count the tiles in order to obtain an approximate area for the whole patio. Archimedes then thought of the leftover regions of irregular shape at the edges as being filled in with smaller tiles, which gave a better fit. His method was to improve these approximations by a limiting process.

However, it was only 300 years ago that Newton and Leibniz brought system to this way of thinking and related integration to differentiation via the Fundamental Theorem, which we shall study in this chapter. In the Examples and Exercises, we will see Riemann sums and trapezoidal sums along with an application to average values.

Problems will define a modified trapezoidal sum, midpoint evaluation, and Simpson's rule. To begin, however, we will examine a most elementary and familiar example, the circle.

EXAMPLE: π AND THE AREA OF A DISC

How is the number π calculated? That is, how do we arrive at such a statement as "π = 3.1415927"? Remember, π is defined in the first place as the ratio of the circumference of a circle to its diameter. We could just measure along the rim of a real disc of metal that had been carefully manufactured to have a diameter of one inch. However, any error in manufacture (perfectly flat disc? perfectly circular rim?) would affect accuracy, so we might be quite doubtful about even 3-digit accuracy in the disc itself. Also, physical measurement of the length of a curved line does not admit of much accuracy; even

2-digit precision would be surprising here. The ancient Babylonians thought that π = 3, and they had tape measures. Evidence of this belief can also be seen in the *Bible* (I Kings 7:23 and II Chronicles 4:2).

It would be equally fruitless to measure the area of that metal disc, which should, of course, be $\pi r^2 = \pi/4$, since the only way to measure physical area is by making linear measurements and conceptually fitting rectangular grids on the region. However, there is a conceptual way to measure area. For arithmetic simplicity, let us take a disc of *radius* 1 (so its diameter is 2) and theoretical area π. Let it be the disc in the plane whose circular rim is the graph of $x^2 + y^2 = 1$ (Figure 7.1). If a slice of this disc has central angle π/2, then its area should be one-fourth the total area, or π/4, because of the "circular" symmetry of the figure. Accordingly, we seek to• *compute theoretically* the area of the region under the graph of

Figure 7.1

$f(x) = \sqrt{1-x^2}$ in the first quadrant (Figure 7.2). Suppose we first divide the interval $[0,1]$ into four pieces $[0, 1/4]$, $[1/4, 1/2]$, $[1/2, 3/4]$, and $[3/4, 1]$. Above each piece we construct the largest rectangle that has that piece for a base and lies inside the region we are measuring. Each rectangle has

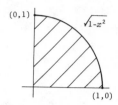

Figure 7.2

width 1/4 and height $f(x)$ for x at the right hand edge (Figure 7.3). The area of these pieces is, respectively,

$$\frac{\sqrt{1 - 1/16}}{4} \ , \ \frac{\sqrt{1 - 1/4}}{4} \ , \ \frac{\sqrt{1 - 9/16}}{4} \ ,$$

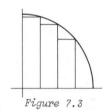

Figure 7.3

and 0, and the sum of these areas is 2.4957091/4. This is a very poor approximation to $\pi/4$.

Next, we construct over the same pieces of the interval $[0,1]$ the smallest rectangles on those pieces as bases that together contain the region in question (Figure 7.4). Their areas are, respectively,

$$\frac{1}{4} \ , \ \frac{\sqrt{1 - 1/16}}{4}, \ \frac{\sqrt{1 - 1/4}}{4}, \ \frac{\sqrt{1 - 9/16}}{4},$$

Figure 7.4

and the sum of these areas is 3.4957091/4, which is the earlier sum plus $\frac{1}{4}$. The average of these two sums is 2.9957091/4; we may thus say *with mathematical certainty* that 2.4957091 < π < 3.4957091 and take the average as our first approximation for π. That average is about 3, and it is hardly a sharp estimate for π. Nevertheless, our scheme of calculation is open to refinement. We next subdivide the base interval $[0,1]$ into

83

ten equal pieces and calculate as before, first the *lower sum* L_{10} of

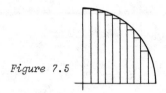

Figure 7.5

areas of rectangles lying inside the region (Figure 7.5). This sum $L_{10} =$

$$\frac{\sqrt{1 - (1/10)^2}}{10} + \frac{\sqrt{1 - (2/10)^2}}{10} + \ldots + \frac{\sqrt{1 - (9/10)^2}}{10} + \frac{\sqrt{1 - (10/10)^2}}{10}.$$

Another way of writing such a sum uses the *sigma notation*

$$L_{10} = \sum_{i=1}^{10} \frac{\sqrt{1 - (i/10)^2}}{10}.$$

In this form some arithmetic manipulation is possible, and we can simplify the actual details of calculation as follows:

$$L_{10} = \sum_{i=1}^{10} \frac{\sqrt{1 - (i/10)^2}}{10}$$

$$= \frac{1}{10} \sum_{i=1}^{10} \sqrt{1 - (i/10)^2}$$

$$= \frac{1}{10} \sum_{i=1}^{10} \sqrt{\frac{100 - i^2}{100}}$$

$$= \frac{1}{10} \sum_{i=1}^{10} \frac{\sqrt{100 - i^2}}{10}$$

$$= \frac{1}{100} \sum_{i=1}^{10} \sqrt{100 - i^2}.$$

We compute $L_{10} = 0.7261295$. Next our *upper sum*:

$$U_{10} = \sum_{i=0}^{9} \frac{\sqrt{1 - \left(\frac{i}{10}\right)^2}}{10} = \frac{1}{100} \sum_{i=0}^{9} \sqrt{100 - i^2}.$$

Clearly this sum differs from L_{10} in the first summand of U_{10} and the last summand of L_{10} (Figure 7.6):

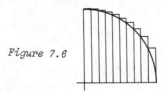

Figure 7.6

$U_{10} = L_{10} + 1/100 \ (\sqrt{100 - 0} - \sqrt{100 - 100}) = L_{10} + 1/10 = 0.8261295.$

Accordingly, we can say $2.9045183 < \pi < 3.3045183$ with assurance, and our average or mean 3.1045183 is now not a poor estimate.

RIEMANN SUMS AND THE INTEGRAL

Our work above suggests a general method for calculating the area A under the graph of a continuous positive function and over an interval of the x-axis. We consider an arbitrary *partition* $p = \{a = x_0 < x_1 < x_2 < \ldots < x_{n-1} < x_n = b\}$ of the interval into subintervals (Figure 7.7). Let l_i = the minimum value of $f(x)$ on the ith interval $\left[x_{i-1}, x_i\right]$, which has length $(x_i - x_{i-1})$, and form a lower sum (Figure 7.8):

$$L_p \sum_{i=1}^{n} (x_i - x_{i-1}) l_i .$$

Just as before, we clearly have $L_p \leq A$. Similarly, if u_i is the maximum value of $f(x)$ on the ith interval of the given partition, then we define the *upper sum* to be (Figure 7.9):

$$U_p = \sum_{i=1}^{n} (x_i - x_{i-1})u_i; \quad A \leqq U_p.$$

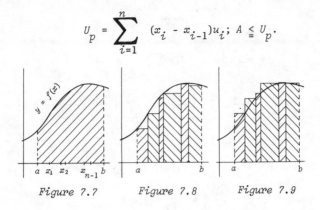

Figure 7.7 Figure 7.8 Figure 7.9

For our purposes, in calculations we will choose partitions p_n of $[a,b]$ into n subintervals of equal length $\frac{b-a}{n}$ and form the sums L_n and U_n. We may expect that by increasing the number n of subintervals sufficiently, we may reduce the maximum error $U_n - L_n$ in the estimation of A to achieve any desired degree of accuracy. In principal, we imagine a limiting value A to which both U_n and L_n tend as n goes toward infinity; this limiting value is defined to be the *area*. There is a special notation for this:

$$A = \int_a^b f(x) \, dx,$$

which is called the *definite integral of* f(x) *on* $[a,b]$.

If our function f were negative instead of positive (Figure 7.10), the sums would all be negative and in the limit they would equal the negative of the area under the (positive) curve $-f(x)$ and over the given interval. Accordingly, if a function f is both positive and negative on an interval $[a,b]$ (Figure 7.11), this process will yield a number

$$\int_a^b f(x) \, dx,$$

86

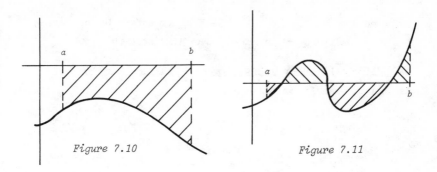

Figure 7.10   Figure 7.11

which is the sum of the areas above the x-axis minus those below the axis. This process of *integration* is also called *quadrature*.

The sums themselves are called *Riemann sums*. The mean $M_n = \frac{L_n + U_n}{2}$ corresponds to the area of the trapezoids in Figure 7.12, which

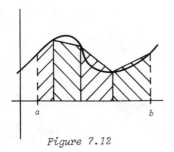

Figure 7.12

approximates the area under curve $y = f(x)$ by the area under the chords. (When l_i and u_i, the extremes of $f(x)$ on the ith subinterval, are not at the two ends of the interval, the picture is not exactly like Figure 7.12.) When f is steadily decreasing on the entire interval $[a,b]$ (as was our example $f(x) = \sqrt{1-x^2}$ on $[0,1]$, the computations are simplified: the minimum on the ith interval is the maximum on the $(i+1)^{st}$ interval when f decreases. Hence $U_n = L_n + (u_1 - l_n)\frac{b-a}{n}$ and the mean $M_n = L_n + (u_1 - l_n)\frac{b-a}{2n}$ is in error

87

by no more than $\left|(u_1-l_n)\frac{b-a}{2n}\right|$. Similar comments hold, of course, for functions that are increasing over the interval.

EXAMPLE: THE AREA UNDER $f(x) = x \sin x$

As another example, we evaluate

$$\int_0^b x \sin x \, dx,$$

where $b = 2.0287578$ is the point on $[0,\pi]$ at which $f(x) = x \sin x$ has a maximum (this fact is established in Problem P1, Ch. 6), and

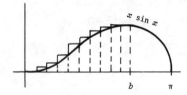

Figure 7.13

f is increasing everywhere on $[0,b]$. The upper sum, then, is (Figure 7.13):

$$U_n = \frac{b}{n} \sum_{i=1}^{n} \frac{ib}{n} \sin\left(\frac{ib}{n}\right)$$

where b/n is the width of each subinterval and ib/n is the right endpoint of the ith interval. For $n = 10$ we calculate this sum to be 1.9785026 and $M_n = U_n + (l_1-u_n)\frac{b}{2n} = U_n - 0.1845871 = 1.7939155$; the maximum error is $\left|(l_1-u_n)\frac{b}{2n}\right| \doteq 0.2$.

88

Average Values

In the process above, we added up the values of the function f computed at 10 evenly spaced points along the interval $[0,b]$ and then multiplied by $\frac{b-0}{10}$; if we had merely divided the sum by 10, we would have computed the average of those values. That is, U_n/b, or even better M_n/b, is an average of the values of $x \sin x$ on the interval $[0,b]$: $M_n/b = 0.8842433$. In general, we define the *average value of a continuous function* f *on an interval* [a,b] to be

$$\frac{1}{b-a} \int_a^b f(x) \ dx.$$

Fundamental Theorems

The first form of the FUNDAMENTAL THEOREM OF THE CALCULUS says that *if we define a new function by the definite integral of* f, *say* F(t) = $\int_a^t$ f(x) dx, *then* F'(t) = f(t). Here is the sketch of a proof phrased in terms of areas. If we think of $F(t)$ as the area under the graph of f and over the x-axis from a out to t, then the rate at which the region adds to its area as t moves to the right is just the height $f(t)$ of the graph at $x = t$ (Figure 7.14).

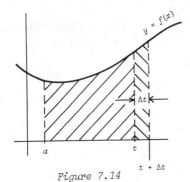

Figure 7.14

The second form of this THEOREM is useful for evaluating definite integrals. It states that *if* G'(t) = f(t) *then* $\int_a^b$ f(t) dt = G(b) - G(a). Any function $G(t)$ for which $G'(t) = f(t)$ is called an *antiderivative* of f, so the process of integration may be reduced to a process of finding and evaluating antiderivatives. For instance, in our example above we calculated

$$\int_0^b x \sin x \, dx \doteq 1.7939155.$$

But if $G(t) = \sin t - t \cos t$, then $G'(t) = t \sin t$, so $\int_0^b x \sin x \, dx = G(b) - G(0) = \sin b - b \cos b = 1.7939112.$

It is remarkable that our calculation was correct through the fifth decimal place! This is much more accuracy than the theory guaranteed. You can understand why M_{10} is a good approximation by noticing that the graph of f bulges upward above the chords about as much as it sags downward below those chords, thus averaging errors. But even more remarkable is the ease with which the Fundamental Theorem gives us a *theoretically exact answer*. The theorem has reduced the problem of evaluating the integral to the problem of calculating the values of the sin and cos functions.

TRAPEZOIDAL SUMS

Our upper and lower sums have been simple to calculate because we have dealt with functions that were increasing (or decreasing) over the entire interval of integration. However, in the general case it would be an insurmountable task to locate the maximal point or the minimal point on each subinterval before evaluating an estimating sum. For practical computations, then, we shall modify our definition of the mean sum over an interval $\left[x_{i-1}, x_i\right]$ to take the average of the values at the two ends of that interval times its width,

$$\frac{f(x_i-1) + f(x_i)}{2} (x_i - x_{i-1})$$

(Figure 7.15). This is again the area of a trapezoid, and if f is increasing (or decreasing) throughout $\left[x_{i-1}, x_i\right]$, this trapezoidal

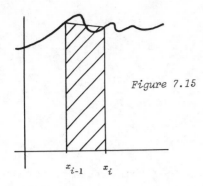

Figure 7.15

area is just one of the summands of the mean sum M_n (Figure 7.16).

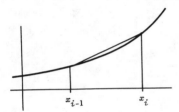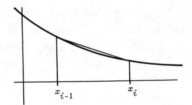

Figure 7.16

The *trapezoidal sum* T_n for $\int_a^b f(x) \, dx$ is thus defined to be

$$T_n = \frac{b-a}{n} \left[\frac{f(x_0) + f(x_1)}{2} + \frac{f(x_1) + f(x_2)}{2} + \ldots + \frac{f(x_{n-1}) + f(x_n)}{2} \right].$$

Here each value $f(x_i)$ shows up twice, divided by two, except for $f(x_0)$ and $f(x_n)$, so

$$T_n = \frac{b-a}{n} \left[\frac{f(x_0) + f(x_n)}{2} + \sum_{i=1}^{n-1} f(x_i) \right].$$

Each mean sum M_n that we have calculated is an example of this trapezoidal sum. This is simple to calculate for any function f,

91

yet $L_n \leq T_n \leq U_n$, so $\lim\limits_{n \to \infty} T_n = A$. In fact, if f has a continuous second derivative that is bounded by K (so $|f''(x)| \leq K$ for every x in $[a,b]$), then the error in T_n as an estimate for $\int_a^b f(x)\ dx$ is no more than

$$\frac{K(b-a)^3}{12n^2}$$

This is called the *truncation error* for T_n; thus the truncation error goes to zero like $1/n^2$. (For a proof of this fact see the book by Courant and John or by James or by Loomis in the Bibliography.)

We emphasize that in each of our examples M_n and T_n are the same (in Exercises 5, 6, and 9 they are different). Throughout this book, our principal method of *numerical quadrature* will be the calculation of trapezoidal sums T_n (but see Problems P1, P2, and P3 for an easily calculated improvement on T_n).

EXAMPLE: THE SINE INTEGRAL

The function $f(x) = x^{-1} \sin x$ has no elementary antiderivative. Hence the *Sine Integral*, the function $Si(x) = \int_0^x t^{-1} \sin t\ dt$ where $\sin 0/0$ is taken to mean 1, must be evaluated by numerical methods. This function is useful in the mathematical analysis of wave propagation. Let us calculate $Si(1)$ by forming the trapezoidal sums T_5 and T_{10}. First we express T_5:

$$T_5 = \frac{1}{5}\left(\frac{1 + \sin 1}{2} + 5 \sin \frac{1}{5} + \frac{5}{2} \sin \frac{2}{5} + \frac{5}{3} \sin \frac{3}{5} + \frac{5}{4} \sin \frac{4}{5}\right)$$

$$= 0.9450788.$$

Next, to express T_{10} we need only add the terms of that sum that are not already in T_5. This gives us

$$T_{10} =$$
$$\frac{1}{10}\left(5T_5 + 10 \sin \frac{1}{10} + \frac{10}{3} \sin \frac{3}{10} + \frac{10}{5} \sin \frac{5}{10} + \frac{10}{7} \sin \frac{7}{10} + \frac{10}{9} \sin \frac{9}{10}\right)$$
$$= 0.9458321.$$

92

The correct value is $Si(1) = 0.9460831$; hence our sums are incorrect in the third decimal place. The sum T_{10} is about five times as accurate as T_5.

EXERCISES

1. Calculate the sums L_4 and U_4 and their mean M_4 for each of the definite integrals here; then use the Fundamental Theorem to evaluate the integral and compare your results.

*a. $\int_1^2 3\,dx$ c. $\int_{-1}^0 (x-x^2)\,dx$ e. $\int_0^{\pi/2} \sin x\,dx$

*b. $\int_0^1 x^3\,dx$ *d. $\int_2^7 \sqrt{x}\,dx$ f. $\int_0^{\pi/6} \dfrac{\cos 2x}{3}\,dx$

*2. Show that in the computation of the area $\int_0^1 \sqrt{1-x^2}\,dx$ of the quarter circle all the trapezoidal sums $T_n = M_n$ must be less than $\pi/4$. How many equal intervals must be used to be sure of the first five decimal places of your answer $4T_n \doteq \pi$?

3. Find L_{20}, $T_{20} = M_{20}$, U_{20}, and then $4T_{20} \doteq \pi$, for $\int_0^1 \sqrt{1-x^2}\,dx$.

*4. Find the area of the region between the graphs of the functions $f(x) = x^2$ and $g(x) = x - 1$ which lies above the interval $[1,2]$. This area may be approximated by the rectangles above intervals of a subdivision of $[1,2]$, where the bottom of the rectangle above a point x will be nearly at $x - 1$ and the top will be nearly x^2. Hence,

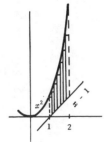
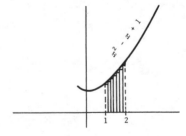

Figure 7.17

this problem is the same as the problem of finding the area under the graph of $f(x) - g(x) = x^2 - x + 1$ above $[1,2]$, which is $\int_1^2 (x^2-x+1)\,dx$. Evaluate this by finding L_5, T_5, and U_5, then L_{10}, T_{10}, U_{10} and

93

compare your results with the correct area $F(2) - F(1)$, where $F'(x) = f(x) - g(x)$.

5. Suppose the average daily temperature in a certain town is found to be $f(t) = 62 + 37 \cos (\frac{\pi t}{6} + 3.1)$ degrees Fahrenheit at a time t months after 31 December. Sketch a graph of this function. What is the average yearly temperature? Calculate the trapezoidal approximation T_5 for the average temperature during the summer (June, July, and August) and compare your results to the correct value, which you can find by use of an antiderivative. This is the first instance where the mean M_5 is not quite the same as the trapezoidal sum T_5. Do you see why from your sketched graph? Also, do you see graphically why $T_5 < A$?

6. Let $v(t) = 100 \sin (\pi t^2)$ be the speed in kilometers per hour of a freight train at a time t hours after leaving Alphaville on its way to Betatown. If the train makes this trip in one hour, find the distance from Alphaville to Betatown.

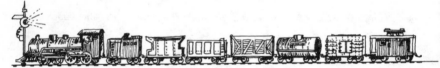

(Hints: leave multiplication by 100 until last. Do sums T_5 and T_{10}, find T_{10} by averaging T_5 with the sum corresponding to the in-between values 0.1, 0.3, 0.5, 0.7, 0.9 of x that were not used to form T_5.) Can you find an antiderivative for $v(t)$?

7. Since the derivative of the function arctan x is $1/(1+x^2)$, the integral

$$\int_0^1 \frac{dx}{1+x^2}$$

is equal to arctan 1 - arctan 0 = $\pi/4$ - 0 = $\pi/4$. Calculate π by evaluating this integral using a tenfold subdivision of [0,1] and a trapezoidal sum. Also, give the percentage of truncation error.

8. Evaluate the trapezoidal sum T_{20} for the integral

$$\int_0^{\pi/2} \sin x \, dx = 1.$$

*9. The function $(1 - \cos t)/t$ has no elementary antiderivative. Hence the function $\text{Cin}(x) = \int_0^x t^{-1} (1 - \cos t) \, dt$ must be evaluated by numerical means. Here $(1 - \cos 0)/0$ means 0. (This integral is related to the *Cosine integral* $\text{Ci}(x) = \gamma + \ln x - \text{Cin}(x)$. It is useful in the study of wave form propagation.)

Estimate $\text{Cin}(0.7)$ by evaluating the trapezoidal sums T_5 and T_{10} for this integral. Compare your results to the correct value, which is 0.1200260.

10. The function $f(\theta) = \cos (a \sin \theta)$ has no elementary antiderivative. Hence the *Bessel function*

$$J_0(x) = \frac{1}{\pi} \int_0^\pi \cos (x \sin \theta) \, d\theta,$$

which is useful in physical applications of math, must be evaluated by numerical methods. Use the trapezoidal sums T_{10} for tenfold subdivisions of the interval $[0,\pi]$ to evaluate $J_0(0.3) = 0.9776262$ and $J_0(2.8) = -0.1850360$.

11. The growth rate for a certain species of fish is known to be $1/\sqrt{17.6t}$ when it is t years old. Express the growth of this fish during its third year of life as an integral. Then estimate that integral by the trapezoidal sum T_5. Also, use the Fundamental Theorem of the Calculus, together with an antiderivative for the growth rate, to evaluate this integral and compare this value with your estimate.

12. A distributor observes that his refrigerator sales rate cycles during the year to grow according to the formula

$$S(t) = 1.012^t \left(\sin \left[\frac{\pi}{12}(t+1.7) \right] + 6.2 \right).$$

Here t is measured in months and $S(t)$ is sales in hundred of units per month. Estimate his total sales during August and September by means of the trapezoidal sum T_5.

PROBLEMS

P1. There is a correction term for trapezoidal sums, so that the *modified sum* for $\int_a^b f(x) \ dx$ is

$$C_n = T_n + \frac{f'(a) - f'(b)}{12} \left(\frac{b-a}{n} \right)^2.$$

Here $\frac{b-a}{n}$ is the length of a single subinterval, the new term involves the values of the derivatives of the integrand at the two endpoints only. It can be shown that if the fourth derivative $f^{(4)}$ exists and is bounded by K on $[a,b]$, then the truncation error in C_n is at most $K(b-a)^5/720n^4$. (See the book by Loomis in the Bibliography. The correction term added to T_n is easily seen to be an estimate of the truncation error for T_n.)

To appreciate the truly remarkable improvement this correction term provides, redo Exercise 4 to find C_5 and C_{10} for $\int_1^2 (x^2-x+1)dx$ and compare your results with the correct area. Then calculate C_5 in Exercise 5, C_5 and C_{10} in Exercise 6, C_{10} in Exercise 7, 9, and 10, and C_{20} in Exercise 8. In each case, observe the number of correct decimal places in your answer and compare to the accuracy of your earlier estimates. Clearly, C_n provides an effective and simple method of highly accurate numerical quadrature.

P2. Apply the refined estimates C_n described in Problem P1 to the Example in the text: that of finding the area of a disc of radius 1. This cannot be done for the integral $\int_0^1 \sqrt{1-x^2} \ dx$ as it stands,

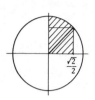

since the derivative of the integral does not exist at $x = 1$: the tangent to the circle is vertical at $x = 1$. Nevertheless, try an altered strategy based on the symmetry of the figure. First find the sum C_{10} for the integral

Figure 7.18

$$\int_0^{\sqrt{2}/2} \sqrt{1-x^2}\ dx.$$

This will provide an estimate of the shaded area in Figure 7.18. But the area of the quarter disc that is omitted by this integral is the same, by symmetry, as the area of the shaded region above the square. Hence, calculate $\pi \doteq 4(2\,C_{10} - 1/2) = 8\,C_{10} - 2$.

P3. In working Problem P1 you may have been struck by the *exact* answer that both C_5 and C_{10} provided for $\int_1^2 (x^2-x+1)\ dx$. Give a proof that, in fact, C_n will be an exact estimate for $\int_a^b f(x)\ dx$ whenever $f(x)$ is a polynomial of degree no greater than three. Your proof will show that this is true no matter what the numbers n, a, and b are.

Do your proof in three stages. In stage one show that C_n is exact for $\int_a^b 1\ dx$ and $\int_a^b x\ dx$ and $\int_a^b x^2\ dx$ and $\int_a^b x^3\ dx$. In stage two demonstrate that if C_n is exact for $\int_a^b f(x)\ dx$ and r is any real number, then the modified sum is also exact for $\int_a^b r\,f(x)\ dx$. In the third stage show that if the modified trapezoidal sums are exact for a given integer n and for both $\int_a^b f(x)\ dx$ and for $\int_a^b g(x)\ dx$, then C_n is exact for $\int_a^b \left[f(x)+g(x)\right]dx$. Then you may argue that, since you have shown each of these three statements to be true, C_n must be exact for every integrand of the form $r_0x^3 + r_1x^2 + r_2x + r_3$ (where each r_i is a real number).

Your proof means that C_n can be regarded as the sum corresponding to a cubic polynomial approximation to the integrand (at least it is if $n \leqq 4$).

P4. The Riemann sum corresponding to *midpoint evaluation* for $\int_a^b f(x)\ dx$ is

$$\frac{b-a}{n} \sum_{i=1}^{n} f\left(\frac{x_i + x_{i-1}}{2}\right).$$

Calculate this sum when $n = 5$ for the integrals of Exercises 4, 5, and 6 and compare your results with those you obtained for trapezoidal sums T_5. (The accuracy of midpoint evaluation is of the same order as that for trapezoidal sums; see Loomis in the Bibliography for details.)

P5. *Simpson's rule* for approximating $\int_a^b f(x)\ dx$ works only with an *even* number n of subintervals. The sum is

$$\frac{b-a}{3n} \left[f(x_0) + 4f(x_1) + 2f(x_2) + 4f(x_3) + \ldots + 4f(x_{n-1}) + f(x_n) \right].$$

This sum may be viewed as 2/3 T_n plus 1/3 of the midpoint evaluation corresponding to the $n/2$ intervals partitioned by $\{x_0 < x_2 < x_4 < \ldots < x_n\}$. The unexpected coefficients in this sum arise as the correct ones with which to estimate the area under the graph of f by the area under quadratic approximating curves. Such an approximation uses one parabolic curve for each pair of subintervals. A proof of this fact is straightforward.

Approximate the integral $\int_0^1 1/(1+x^2)\ dx$ of Exercise 7 by use of Simpson's rule when $n = 10$. Compare your answer to T_{10}, C_{10}, and also to the midpoint evaluation corresponding to $n = 5$, by choosing the most accurate of these estimates for $\pi/4$ and expressing the error of each estimate as a multiple of the smallest error. (The truncation error for Simpson's rule is of the same order as that

98

for the modified trapezoidal sums. See Courant and John, James, or
Loomis in the Bibliography for details. Simpson's rule provides an
accurate alternative to the modified trapezoidal sum C_n whenever
the derivatives $f'(a)$ and $f'(b)$ are not easily calculated for use
in C_n.)

P6. Describe at least one plausible situation in a field of your
own current interest, perhaps biology or business or chemistry,
where the definite integral may be applied to obtain a useful numerical solution. Discover such a real-life situation by surveying a
current issue of an appropriate journal in your field. (See the
Bibliography for some suggested journal titles.)

Answers to Starred Exercises

Exercises 1a. $L_4 = M_4 = U_4 = \int_1^2 3dx = 3.$

     1b. $L_4 = 0.1406250$

         $M_4 = 0.2656250$

         $U_4 = 0.3906250$

         $\int_0^1 x^3 dx = 0.25$

     1d. $L_4 = 9.6702816$

         $M_4 = 10.439993$

         $U_4 = 11.209704$

         $\int_2^7 \sqrt{x} \, dx = 10.461221$

     2. $n = 10^5$

     4. $L_5 = 1.64$ $T_5 = 1.84$ $U_5 = 2.04$ $L_{10} = 1.735$

       $T_{10} = 1.835$ $U_{10} = 1.935$ $A = 1.8333333$

     9. $T_5 = 0.1199286$ $T_{10} = 0.1200017$

8

LOGARITHMS AND EXPONENTIALS

INTRODUCTION

Nicolas Chuquet first noticed in 1484 that to multiply any two members of the geometric series 1, r, r^2, r^3, r^4, ..., we only need to add their exponents, $r^a \times r^b = r^{a+b}$. Similarly, Chuquet found that division among terms corresponds to subtraction of exponents: $r^a \div r^b = r^{a-b}$. More than 100 years later John Napier made this idea useful by calculating "logarithms" for all 8-digit decimal fractions. It is difficult to overestimate the value of Napier's log tables. They have been used billions of times each year to do accurate multiplications of every conceivable sort in navigation, engineering, science, and business.

We begin this chapter with a definition of the (natural) logarithm function as an integral and illustrate it with the calculation of ln 2. Then we will define the inverse function, the exponential, and construct graphs of ln x and e^x. We will also define the number e in a natural way and calculate it as a limit.

Next we apply our math to the economic concept of compound

interest and to the biophysical operation of carbon dating. These
applications are explored further in the Exercises, along with the
probability integral, Newton's law of cooling, and Huxley's differential growth ratio for the huge claw of fiddler crabs. In the Problems we will discuss the computation of monthly payments on home
loans and evaluate continued fraction expressions for the functions
$\ln x$ and e^x.

THE DEFINITION OF LOGARITHM

We shall use the insights that we have gained thus far in our study
of the calculus to guide us in a new approach to logarithms. These
insights were not available to Chuquet and Napier. This approach
may seem strange to you at first, but it has great mathematical
power. Our method will be to show first that any function (like the
logarithm) that turns multiplication into addition must have a certain kind of derivative. We then appeal to the Fundamental Theorem
to define the logarithm as the integral of its own derivative.

Let us seek a function f that has the property that $f(xy) =$
$f(x) + f(y)$. Clearly $f(1)$ must be zero since $f(y) = f(1y) =$
$f(1) + f(y)$. Then $0 = f(1) = f(x \times \frac{1}{x}) = f(x) + f(1/x)$ so $f(1/x) =$
$-f(x)$. If we differentiate f at a point a we get

$$
\begin{aligned}
f'(a) &= \lim_{h \to 0} \frac{f(a+h) - f(a)}{h} \\
&= \lim_{h \to 0} \frac{f(a+h) + f(1/a)}{h} \\
&= \lim_{h \to 0} \frac{f(\frac{a+h}{a})}{h} \\
&= \lim_{h \to 0} \frac{f(1 + h/a)}{h} \, .
\end{aligned}
$$

If $a \neq 0$, then ah tends to zero just as h does, and we may rewrite
the last limit as

$$= \lim_{ah \to 0} \frac{f(1 + ah/a)}{ah}$$

$$= \frac{1}{a} \left[\lim_{h \to 0} \frac{f(1+h)}{h} \right]$$

$$= \frac{1}{a} f'(1).$$

We have just shown that if f carries products to sums, then the derivative of f at a point a is completely determined by the derivative $f'(1)$ of f at the single point 1: $f'(a) = f'(1)/a$. Now suppose that our function f has a nonzero derivative $f'(1)$ at 1, and we define a new function $F(x) = \frac{1}{f'(1)} f(x)$. We still have

$$F(xy) = \frac{1}{f'(1)} f(xy) = \frac{1}{f'(1)} [f(x)+f(y)] = F(x) + F(y). \text{ Likewise,}$$

$F(1) = 0$. In addition, $F'(1) = \frac{f'(1)}{f'(1)} = 1$, and in general $F'(x) = 1/x$.

By the Fundamental Theorem of the Calculus, the integral

$$\int_1^x \frac{dt}{t} = F(x) - F(1) = F(x).$$

But now we are no longer supposing; we have the function $F(x)$. It is the integral. The functions f and F cannot be defined for the argument 0, since if they were, then $F(0) = F(0 \times 2) = F(0) + F(2)$, which says that $F(2) = 0$. The integral $\int_1^2 1/t \, dt$ is clearly positive, though, since it is the area under $1/t$ and over $[1,2]$ (Figure 8.1). Thus we cannot define this integral for $x \leq 0$. However, for positive x we do have a function $F(x)$ defined by this

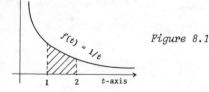

Figure 8.1

102

integral (remember, if $0 < x < 1$, then $\int_1^x 1/t \; dt = -\int_x^1 1/t \; dt$).
It is important enough to merit a special name, the *natural logarithm function*, and special notation:

$$\ln x = \int_1^x \frac{dt}{t}, \; x > 0.$$

EXAMPLE: $\ln 2$

We calculate $\ln 2$ by estimating this integral. The lower sum
$L_{10} = 1/10 \; (1/1.1 + 1/1.2 + \ldots + 1/2.0) = 0.6687714$, the upper sum
$U_{10} = 1/10 \; (1 + 1/1.1 + \ldots + 1/1.9) = 0.7187714$, and the trapezoidal
sum $T_{10} = \frac{L_{10} + U_{10}}{2} = 0.6937714$. The correct value is $\ln (2) = 0.6931472$, so T_{10} is accurate to (nearly) three decimal places; its
error is 0.0006.

THE GRAPH OF $\ln x$

We can graph the natural logarithm function: $\ln(1) = 0$, and its
derivative $1/x$ is always positive, so $\ln x$ is always increasing

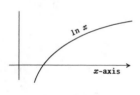

Figure 8.2

(Figure 8.2). Since $1/x$ gets smaller
as x increases, the slope of the graph
becomes less and less for increasing
x. Since $\ln 1/x = -\ln x$, its values
become large-negative for $x < 1$ and
decrease toward $- \infty$ as x decreases
toward 0.

EXPONENTIALS

The logarithm of 2 is greater than one-half. Hence $\ln 4 = \ln(2^2) = 2 \ln 2$ is greater than one: $\ln 4 > 1$. From this fact you can see
that the values of the function $\ln$ eventually grow larger than any
integer N, no matter how big N is. This is true because $\ln(4^N) = N \ln 4 > N$, so that 4^N is an integer whose logarithm is bigger than
N. Since $\ln(4^{-N}) = -\ln(4^N) < -N$, this function takes on values less
than $-N$ as well. Since $\ln x$ is an increasing and continuous

103

function (why is this true?), it takes on every real number as its value precisely once. That is, for each real number y there exists exactly one number x with $y = \ln x$.

Consequently, the function $\ln x$ possesses an inverse function, which we temporarily denote by $\exp(y)$: $\ln(\exp y) = y$ and $\exp(\ln x) = x$, so $y = \ln x$ if and only if $x = \exp y$. This function $\exp$ is defined for every real number y. Its values are the positive real numbers x that are admissible arguments for $\ln x$. Since $\ln (1) = 0$, we have $\exp(0) = 1$. It is also apparent that $\exp(x+y) = \exp(x)\exp(y)$. The slope of the graph of $\exp$ at $y = 0$ is the reciprocal of the slope for $\ln(x)$ at $x = 1$, which is $\ln'(1) = 1$, so $\exp'(0) = 1$. In general, the graph of $\exp(x)$ may be constructed from the graph of $\ln(x)$ by reflection in the line $y = x$ (Figure 8.3).

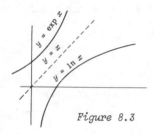

Figure 8.3

The chain rule says that

$$\frac{d}{dy} y = \frac{d}{dy} \ln \exp(y) = [1/\exp y]\exp'(y)$$

so $\exp'(y) = \exp(y)$, which is a surprise!

If a is a positive number and $x = p/q$ is rational, then $\ln(a^x) = x \ln a$ and so $a^x = \exp(\ln a^x) = \exp(x \ln a)$; we take this as a definition of a^x when x is not rational as well. It is easy to verify that $a^{x+y} = a^x a^y$, $(a^x)^y = a^{xy}$, $a^{-x} = 1/a^x$ and $a^x b^x = (ab)^x$ (*laws of exponents*). Notice that if $\ln a = 1$, then $a^x = \exp(x \ln a) = \exp(x)$: there is a special number e such that $\ln e = 1$ or $e = \exp(1)$.

Since $\ln 2 < 1 < \ln 3$, $2 < e < 3$. We can evaluate the integrals by use of trapezoidal sums T_{10} to find that $\ln(2.7) < 1 < \ln(2.8)$; the number e is between 2.7 and 2.8. By definition, $e^x = \exp(x \ln e) = \exp(x)$; we may henceforth use the notation e^x as well as $\exp(x)$ for the *exponential function*.

EXAMPLE: A CALCULATION OF e

We know that the derivative of $\ln x$ at $x = 1$ is $1/x = 1$; hence $1 = \lim_{h\to 0} \dfrac{\ln(1+h) - \ln(1)}{h} = \lim_{h\to 0} \dfrac{\ln(1+h)}{h}$. If we evaluate the exponential function at both sides of this equation we get

$$e = e^1 = \lim_{h\to 0} \exp \frac{\ln(1+h)}{h}$$
$$= \lim_{h\to 0} (1+h)^{1/h}.$$

This limit expresses the number e without mentioning the logarithmic or exponential functions; it may be restated by substituting $1/n$ for h:

$$e = \lim_{n\to\infty} (1 + 1/n)^n.$$

It is fun to evaluate this limit (see Table 8.1).

TABLE 8.1

n	$(1 + 1/n)^n$
10	2.5937425
10^2	2.7048138
10^3	
10^4	
10^5	
10^6	2.7182818

105

EXAMPLE: COMPOUND INTEREST AND GROWTH

Suppose a bank pays interest on deposits at the rate of 10% per year so that $100 deposited will become $110 at the end of a year. But suppose another bank also offers 10% interest, and it agrees to *compound* this *interest* every six months; that is, this bank will pay

5% interest at the end of six months and another 5% interest at the end of the year. After six months in this second bank, the $100 deposit becomes $105, and during the second six months the added $5 earns interest along with the original dollars, yielding $(1.05)(\$105) = \$110.25$. The extra 25 cents is the interest on the $5 interest that was added to the account at midyear.

If interest had been compounded quarterly (that is, every 3 months), then successive quarters' balances would be $\$(1.025)100$, $(1.025)^2 100$, $(1.025)^3 100$, and finally $(1.025)^4 100 = \$110.38129$. If interest had been compounded monthly, at the end of the year there would be a balance of $(1 + 0.1/12)^{12} \times \$100 = \$110.47131$. We record our results in Table 8.2.

TABLE 8.2

Compounding	Year-end Balance
annually	$110.00
semiannually	110.25
quarterly	110.38129
monthly	110.47131
daily	
hourly	110.51745[†]

[†]Round off error has resulted in our hourly-compounding rate being larger than $e^{0.1} = 1.1051709$; the correct value is 1.1051703, which is less than $e^{0.1}$.

106

The limiting number is of course $\left[\lim\limits_{n \to \infty} (1 + \frac{0.1}{n})^n\right]$ times the \$100 initial deposit. Since $10n = m$ goes toward infinity as $n \to \infty$, we have

$$\lim_{n \to \infty} (1 + 0.1/n)^n = \lim_{m \to \infty} (1 + 1/m)^{m/10}$$

$$= \lim_{m \to \infty} \left[(1 + 1/m)^m\right]^{0.1}$$

$$= \left[\lim_{m \to \infty} (1 + 1/m)^m\right]^{0.1}$$

$$= e^{0.1} = 1.1051709.$$

That is, if 10% interest were *compounded continuously*, it would pay 10.51709% yearly (Figure 8.4). Be sure you understand the difference between a *nominal rate* of 10% per year and an *effective rate* of 10.51709 % after a year's time.

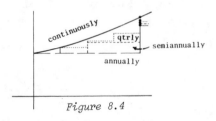

Figure 8.4

EXAMPLE: CARBON DATING AND DECAY

A radioactive element like carbon 14 decays at a rate proportional to the amount present. That is, the decomposition of an individual atom of C^{14} is a random event, but in any amount of observable size (that is, an amount containing many millions of atoms), half of the atoms present today will have decomposed in 5,568 years. Now there is a certain stable amount of C^{14} in the carbon dioxide we breathe; it is continuously replenished by the irradiation of the upper atmosphere. Trees use some of this carbon dioxide to make their wood. When a tree dies, however, it can no longer consume carbon

dioxide, and the C^{14} that the tree accumulated while alive will be half gone in 5,568 years. This fact has given scientists a means of dating archaelogical objects. As a sample problem, suppose that an axe handle has 1/5 the proportion of C^{14} that new wood has today. How old is it?

We know that the rate of change of the quantity $y(t)$ of C^{14} at times t is $y'(t) = k\,y(t)$ for some constant k of proportionality. But we can guess at a function $y(t)$ that behaves this way: if $y(t) = e^{kt}$, then $y'(t) = k\,e^{kt} = k\,y(t)$. Furthermore, the same is true if we add any other constant c to the exponent to get $y(t) = e^{kt+c} = e^c e^{kt}$. Thus if the amount of C^{14} in an equal weight of new wood is e^c (corresponding to time $t = 0$), then 5,568 years later it will be $e^c/2 = e^c e^{5568k}$, and we have

$$e^{5568k} = 1/2$$

$$5568k = \ln 1/2$$

$$k = (\ln 1/2)/5568$$

$$k = -0.0001245.$$

Now the axe handle is observed to have only 1/5 of its C^{14} remaining, so its age is determined by (see Figure 8.4):

108

$$e^{kt} = 1/5$$

$$kt = \ln 1/5$$

$$t = (\ln 1/5)/k$$

$$t = 12928.495 \text{ years.}$$

EXERCISES

1. In each case estimate the natural logarithm by calculating the trapezoidal sum T_5 for the integral that defines the logarithm:

*a. ln 3   *b. ln 9   *c. ln 4   d. ln 4.5   *e. ln 1.01   f. ln 16

2. Estimate ln 2.7 and ln 2.8 by calculating the sums T_{10} for each integral; thus show that $2.7 < e < 2.8$.

*3. The derivative $\dfrac{d}{dx} a^x = \dfrac{d}{dx} e^{x \ln a} = \ln a \, e^{x \ln a} = (\ln a)a^x$; accordingly, the derivative of 2^x at $x = 0$ is ln 2:

$$\ln 2 = \lim_{h \to 0} \frac{2^h - 1}{h}.$$

By taking repeated square roots, evaluate the stages in this limit for $h = 1, 1/2, 1/4, 1/8, 1/16, 1/32$, and record your results in a table. The correct limit is ln 2 = 0.6931472; improve the accuracy of this method by calculating the slopes of chords whose center is at $x = 0$, instead of chords with one end at $x = 0$:

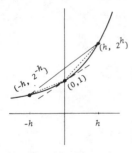

Figure 8.5

$$\ln 2 = \lim_{h \to 0} \frac{2^h - 2^{-h}}{2h} .$$

Tabulate these slopes for the same values of h as above. Next, find out what power of ½ will suffice for h in order to achieve an accuracy of 6 decimal places. Does this provide a reasonably

109

accurate method for computing logarithms on a calculator with a square-root button but no ln function?

*4. Use the method of Exercise 3, taking $h = 1/32$ to estimate ln 2.7 and ln 2.8. Then interpolate to estimate the number e for which ln $e = 1$; that is, find the number a so that

$$a(\ln 2.8 - \ln 2.7) + \ln 2.7 = 1,$$

then estimate $e \doteq 2.7 + (2.8-2.7)a$. This process is a numerical inversion of the function ln to find e^x at $x = 1$; your result is surprisingly close, isn't it?

5. Find the rate of continuously compounded interest that will yield 10% per year: thus, find x so that $e^{xt} = 1.1$ when $t = 1$; $x = \ln 1.1$. Compute x using the method developed in Exercise 3.

*6. The *present value* PV and *future value* FV of a sum of money are related by the interest rate I on an annual basis, the time N in years, and the frequency φ, which is the number of intervals per year with which the interest is to be compounded. The formula is $FV = PV(1 + I/\varphi)^{\varphi N}$. Calculate the value of $1000 after 10 years with interest at an annual rate of 0.1075 (which is 10-3/4%) compounded annually, quarterly, and monthly. Then invent a formula relating present and future values when interest is compounded continuously. Give a limiting argument that your formula is correct and apply your formula to the case above with an annual rate of 0.1075.

*7. Continue the investigation of Exercise 6 by deriving a formula for computing the interest rate I when you know the present and future values, the term of time N years, and the frequency φ per year of compounding. Then compute the interest rate required to double your money in 5 years if interest is compounded annually, quarterly, and monthly. Next, invent a formula to find the appropriate rate when interest is continuously compounded. Give a limiting argument that your formula is correct and apply it to find the continuous rate that doubles the value in 5 years.

110

*8. The *error function* or *probability integral*

$$\text{erf } x = H(x) = \frac{2}{\sqrt{\pi}} \int_0^x e^{-t^2}\, dt$$

must be evaluated by numerical quadrature since e^{-t^2} has no elementary antiderivative. Find erf(1.23) by use of the trapezoidal sums T_{10} and T_{20}.

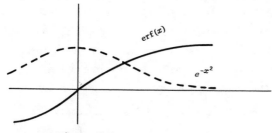

Figure 8.6

*9. The *half-life* of radium, the length of time during which ½ of a given mass of radium will decompose, is about 1622 years. If a hospital has some radium in a pellet, how long may the pellet be used with the assurance that its radiation level has changed by no more than 2%?

10. The population of a bacterial culture is observed to increase by 9% in an hour and 10 minutes. What is the population increase over 24 hours (assume that the growth is exponential)?

11. Newton's *Law of Cooling* says that the rate of cooling of a heated object will be proportional to the difference of its temperature and the temperature of its environment. If a thermometer in a pottery kiln registers 1400° C when the heat is turned off and

and then reads 1300° C in 23 minutes, how many hours will be required for the kiln to cool down to 50° C (assume an outside temperature of 20° C)?

12. Suppose \$100 was invested at an annual interest rate I *without compounding;* find the time $Y = Y(I)$ in years required to double the original sum. Next, find the total sum after Y years at the annual rate $I = 0.12$ if interest were compounded every Y/n years for values of n = 2, 10, 100, 1000, 10000. Finally, guess at the limiting value of the investment as $n \to \infty$ and prove your guess is correct.

13. The male fiddler crab (*Uca minax*) in its immediately post-larval stage has two claws (chelae) of equal size. Each claw weights roughly 1.2 mg, which is about 2% of the weight (60 mg) of the rest of the body. However, a mature specimen is grotesque: one of its claws is disproportionately large. In one case this huge claw weighed 7.25 g, which amounted to 72% of the rest-of-body weight of 10.06 g! Sir Julian Huxley reasoned as follows about these facts. Let y stand for the size of the organ (chela) and x that of the rest of the body and assume that both x and y satisfy growth equations of the form

$$x(t) = ae^{\alpha Gt} \qquad y(t) = be^{\beta Gt}.$$

Here a and α or b and β are specific constants for the rest-of-body or for the organ-in-question and G measures general conditions of growth as affected by age and environment.

Show that these assumptions imply that there are constants c and k for which

$$y = cx^k.$$

Huxley called k the *differential growth ratio*; he emphasized that k is a constant independent of age and environmental factors.

Next, use the data given above to establish the constants c and k for the case in question. Then find the total weight of a fiddler crab at the stage where his larger claw has grown to one-half the weight of the rest of his body.

14. The atmospheric pressure at a height h above a planet's surface is given by an exponential function:

$$p(h) = ce^{-kh}.$$

Suppose on a given day the sea level pressure is 765 mm of mercury, and at the top of a 152 meter hill the pressure is 747 mm on your barometer. What is the pressure at the top of Mt. Everest (8848 m)? What is the altitude at which the pressure is ½ that at sea level? 1/4 of that at sea level? 1/8 of that at sea level?

PROBLEMS

*P1. Use the modified trapezoidal sum $C_5 = T_5 + \dfrac{f'(a)-f'(b)}{12}\left(\dfrac{b-a}{n}\right)^2$

(see Problem P1, Ch. 7) to estimate ln 4 in Exercise 1c. Then compute C_{10} and C_{20} for the integral of Exercise 8. In each case, discuss the improvement in your estimate over the trapezoidal sum by calculating the error of C_n and comparing it to the error of the corresponding T_n (ln 4 = 1.3862944 and erf(1.23) = 0.9180501).

113

P2. Express the appropriate monthly payment PMT due on a loan of PV dollars for n months (or N years) if interest at a monthly rate i (or an annual rate I) is applied each month to the unpaid balance ($n = 12N$ and $i = I/12$). To do this, imagine that the monthly payments are paid into a separate account that earns interest monthly on its balance. Then the separate account will have $PMT(1+i)^{n-1}$ in it for the payment made at the end of the first month together with its compound interest for $n-1$ months. There will also be added $PMT(1+i)^{n-2}$ for the second payment, and so on, to total

$$PMT\left[(1+i)^{n-1} + (1+i)^{n-2} + \ldots + 1\right].$$

This geometric series has the sum

$$PMT\,\frac{1-(1+i)^n}{1-(1+i)} = PMT\,\frac{(1+i)^n-1}{i}\ .$$

(Remember that if $x \neq 1$, then $1 + x + x^2 + \ldots + x^{n-1} = (1-x^n)(1-x)$). This amount in the special account must equal the future value $PV(1+i)^n$ of the amount borrowed; this gives $PMT = PVi/\left[1-(1+i)^{-n}\right]$.

Compute the payments due on a home loan of \$31,300 for 25 years at 9-1/8% annual rate. If you could only pay \$300 per month, how much money could you borrow? If you paid \$300 per month for a 25-year loan of \$31,300, what would the interest rate be (solve iteratively)?

Next, take a limit to derive the recipe $PMT = PV\,\frac{I}{N}\,(1-e^{-IN})$ for the same situation except that interest is compounded continuously and the payments are to be made continuously. Compute with this recipe the payments due on \$31,300 for 25 years at 9-1/8% annual rate. How good is this continuous model as an approximation to the exact payment calculated above?

P3. Leonhard Euler (1707-1783) offered the following continued fraction expansion (see Problem P5, Ch. 6):

$$\frac{e-1}{2} = \frac{1}{1+}\ \frac{1}{6+}\ \frac{1}{10+}\ \frac{1}{14+}\ \frac{1}{18+}\ \cdots\ .$$

Another is

$$e = \frac{1}{1-} \ \frac{1}{1+} \ \frac{1}{2-} \ \frac{1}{3+} \ \frac{1}{2-} \ \frac{1}{5+} \ \frac{1}{2-} \ \cdots \ .$$

Find the number of terms of each expansion necessary to compute e correctly to 5 decimal places.

The second of these expansions may be generalized to compute values of the exponential function:

$$e^x = \frac{1}{1-} \ \frac{x}{1+} \ \frac{x}{2-} \ \frac{x}{3+} \ \frac{x}{2-} \ \frac{x}{5+} \ \frac{x}{2-} \ \cdots \ .$$

Two other continued fraction expansions for this function are

$$e^x = 1 + \frac{x}{1-} \ \frac{x}{2+} \ \frac{x}{3-} \ \frac{x}{2+} \ \frac{x}{5-} \ \frac{x}{2+} \ \cdots$$

and

$$e^x = 1 + \frac{x}{1-x/2+} \ \frac{x^2/(4 \mathrm{X} 3)}{1+} \ \frac{x^2/(4 \mathrm{X} 15)}{1+} \ \frac{x^2/(4 \mathrm{X} 35)}{1+} \ \cdots \ \frac{x^2/[4 \mathrm{X} (4n^2-1)]}{1+} \ \cdots$$

Use the first 6 terms of each of these expressions to compute $e^{1.23}$; compare your results to the correct figure $e^{1.23} = 3.4212295$. (It is clear that the last expression offers a highly efficient computational method.)

Next, calculate ln 3.4212295 using the continued fraction expansion

$$\ln(1+x) = \frac{x}{1+} \ \frac{x}{2+} \ \frac{x}{3+} \ \frac{4x}{4+} \ \frac{4x}{5+} \ \frac{9x}{6+} \ \cdots \ .$$

P4. Write out the first five terms of the binomial expansion of $\left(1 + \dfrac{x}{n} \right)^n$. Now take the limit of each of these terms as n goes to

115

infinity. This results in a polynomial of degree four in x.
Evaluate this polynomial for $x = 0.12345$ and compare the result
to e^x. Is this an efficient way to compute values of e^x on a machine without a button for that function?

P5. Describe at least one plausible situation in a field of your
own current interest, perhaps biology or business or chemistry,
where the logarithmic and exponential functions may be applied to
obtain a useful numerical solution. Discover such a real-life
situation by surveying a current issue of an appropriate journal in
your field. (See the Bibliography for some suggested journal titles.)

Answers to Starred Exercises and Problems

Exercises 1a. 1.1102675   1c. 1.4134836
           1b. 2.3773042   1e. 0.0099503
            3. for $h = 1/32$, $(2^h-1)/h = 0.7007088$ and
                $(2^h-2^{-h})/2h = 0.6932014$; for
                $h = 1/512$, $(2^h-2^{-h})/2h = 0.6931474$
            4. $\ln 2.7 \doteq 0.9934113$ $\ln 2.8 \doteq 1.0297971$
                     $e \doteq 2.7181080$
            6. annual compounding yields $2776.1143;
                continuous compounding yields $FV = PVe^{IN}$
                where I is the rate per year over N years

            7. $I = \varphi\left[\left(\dfrac{FV}{PV}\right)^{\frac{1}{\varphi n}} - 1\right]$; the annual rate is 0.1486984;

                the continuous rate is 0.1386294
            8. $T_{10} = 0.9172793$, $T_{20} = 0.9178574$;
                $\mathrm{erf}(1.23) = 0.9180501$
            9. 47.275373 years

Problems   P1. $C_5 = 1.3853572$;
                $C_{10} = 0.9179624$
                $C_{20} = 0.9180282$

116

9

VOLUMES

INTRODUCTION

In his *Mathematical Thought from Ancient to Modern Times* (see Bibliography) Morris Kline describes the four major types of problems that led to the creation of the calculus. These are the problems of calculating:

(i)   Motion: The distance traveled by an object, its velocity, its acceleration, and the relationships between these quantities.

(ii) Tangents: The tangents to curves and to solids. These were needed in geometry, and they were also useful in optics and in the study of motion.

(iii) Extremes: Maxima and minima for various functions. Notably, Galileo (1564-1642) found the correct angle ($\pi/4$) to fire a cannon for maximal range.

(iv)   Lengths, areas, and volumes: The distance traveled by a
        planet around the sun, areas bounded by curves, volumes
        bounded by curved surfaces, centers of gravity, and total
        gravitational attraction of a body.

In the present chapter we shall study some methods that the calculus
defines for expressing volumes as integrals of functions of real
numbers.

The first significant progress in the development of these
methods was made in the third century B.C. by Archimedes. In order
to calculate the total volume he used "methods of exhaustion" for
conceptually packing small regular shapes into a larger space having
a curved surface. His techniques were special, though, for each
problem, and he was able to obtain answers for only a few special
shapes.

The next important additions to these methods came in the seventeenth century A.D. when Johannes Kepler offered to help wine dealers
find the volumes of their kegs. Kepler found the volume of a ball by
considering it to be made up of many cones of various sizes, all having their vertices at the ball's center. Then he established the
volume of a cone by imagining it sliced into many very thin wafers,
just as we shall do below. Thus he found the cone to have a volume
equal to one-third its height times its base area. Since the ball
was made up of many cones, all with their bases in its spherical surface, the volume of the ball was found to be one-third the radius
times the surface area. (Do you suppose wine was stored in balls
and cones?)

The calculus has since given us several techniques that may be
applied with minimal ingenuity to a wide variety of solid figures.
We shall first give an Example of Kepler's trick, the "slab method,"
for finding the volume of a cone. Another Example of this method
establishes the volume of a ball directly. A final Example uses an
alternative, the "shell method," to develop the volume of the cone.
In each of these Examples a numerical approximation is found to illustrate the method.

These methods are then applied in the Exercises and Problems to find the volumes of many different solids. In each case a numerical sum is calculated to realize a finite approximation, and this sum is compared with a theoretical result obtained by means of the Fundamental Theorem. A final Problem considers the pressure exerted on an underwater viewing porthole at Marineland.

EXAMPLE: THE SLAB METHOD FOR A CONE

Let us find the volume of a cone, a right circular cone of radius r = 1.2 and height h = 3.4 (Figure 9.1). We first rearrange the

Figure 9.1

problem to consider the cone as lying in a space with coordinates x, y, and z; we have the vertex at the origin, and the axis of the cone lies along the x-axis. The line in the (x,z)-plane that goes through the origin and is r = 1.2 units above the x-axis at a distance h = 3.4 units out along the

x-axis is a generator of the cone (Figure 9.2). This line has slope r/h = 1.2/3.4, so its equation is $z = \frac{r}{h} x$ = 0.3529412 x (and y = 0). We now imagine that the region in the (x,y)-plane below the graph of

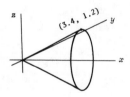

Figure 9.2

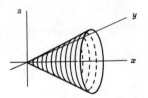

Figure 9.3

this line and above the interval $[0,h]$ of the x-axis is revolved around the x-axis to sweep out the volume of the cone. To find that volume we subdivide the interval $[0, 3.4]$ of the x-axis into n = 10 equal pieces and imagine slicing the cone into 10 slabs with all the faces of the slabs parallel to the (y,z)-plane (Figure 9.3). That is, each slice is perpendicular to the x-axis.

119

A single slice corresponding to the ith subinterval has thickness $h/n = 0.34$. The radius of its larger face is $\frac{r}{h} x_i = 0.3529412\, x_i$, and the radius of its smaller face is $\frac{r}{h} x_{i-1}$ (Figure 9.4). Hence the area of the larger face is $\pi \frac{r^2}{h^2} x_i{}^2 = 0.3913403\, x_i{}^2$ and that of

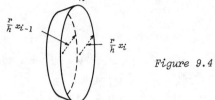

Figure 9.4

the smaller is $\pi \frac{r^2}{h^2} x_{i-1}^2$. The volume of this slab will lie between

$$\pi \frac{r^2}{h^2} x_{i-1}^2 \left(\frac{h}{n}\right) = \frac{\pi r^2 x_{i-1}^2}{hn}$$

and

$$\frac{\pi r^2 x_i^2}{hn}.$$

The total volume of all the slabs lies between

$$\sum_{i=1}^{n} \frac{\pi r^2 x_{i-1}^2}{hn}$$

and

$$\sum_{i=1}^{n} \frac{\pi r^2 x_i{}^2}{hn}$$

120

These two sums, however, are simply the lower and upper sums L_n and U_n for the integral $\int_0^h \frac{\pi r^2}{h^2} x^2 dx$. Since $\frac{\pi r^2 x^3}{3h^2}$ is an antiderivative for the integrand, this definite integral has value $\frac{\pi r^2 h}{3}$, which is the volume of the cone. We estimate this volume for our cone, with radius 1.2 and height 3.4, by calculating for ten slabs the sums L_{10}, U_{10}, and T_{10}. Since $\frac{\pi 1.2^2}{3.4 \times 10} = 0.1330557$, we have

$L_{10} = 0.1330557 (0. + 0.34^2 + 0.68^2 + \ldots + 3.06^2) = 4.3836527;$

$U_{10} = L_{10} - 0. + 0.1330557(3.4)^2 = 5.9217765;$

$T_{10} = \frac{1}{2}(L_{10} + U_{10}) = 5.1527146.$

These results compare with the theoretical volume $5.1270792 = \frac{\pi(1.2)^2(3.4)}{3}$. Doesn't the theoretical method have ease and power?

EXAMPLE: THE SLAB METHOD FOR A BALL

We now apply the same sort of reasoning to a ball of radius r. By symmetry we may estimate the volume of a half-ball and multiply that by 2. Again we place the half-ball in (x,y,z)-space, with the cut face lying in the (y,z)-plane. We subdivide the interval $[0,r]$ into

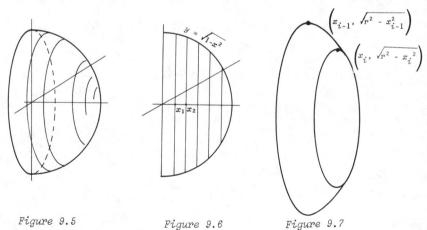

Figure 9.5   Figure 9.6   Figure 9.7

n equal pieces and again slice this solid at each of the points x_i that partition $[0,r]$ (Figure 9.5). The volume of the half-ball is swept out as the region in the (x,z)-plane under the graph of $\sqrt{r^2-x^2}$ and above $[0,r]$ is revolved about the x-axis (Figure 9.6). One slab of this total volume is swept out by the region under the graph of $\sqrt{r^2-x^2}$ and above $[r_{i-1}, r_i]$ (Figure 9.7). This slab has area on the back face of $\pi(r^2 - x_{i-1}^2)$ and area on the front of $\pi(r^2 - x_i^2)$. Its thickness is r/n, so its volume is surely between $\frac{r\pi}{n}(r^2 - x_{i-1}^2)$ and $\frac{r\pi}{n}(r^2 - x_i^2)$.

As in the case of the cone, we may add up similar estimates for each of the n slabs to get a total volume for the half-ball between

$$\sum_{i=0}^{n-1} \frac{r\pi}{n}(r^2 - x_i^2)$$

and

$$\sum_{i=1}^{n} \frac{r\pi}{n}(r^2 - x_i^2).$$

And, again, these estimating sums are, respectively, the upper and lower sums U_n and L_n for an integral, namely $\int_0^r \pi(r^2-x^2)dx$. An antiderivative in this case is $\pi(r^2x - x^3/3)$, so

$$\int_0^r \pi(r^2-x^2)dx = \pi(r^3 - r^3/3) = \frac{2\pi r^3}{3}.$$

(The whole ball has twice this volume or $4\pi r^3/3$.)

If we estimate this integral for a ball of radius r = 2.3 by using n = 10 subintervals to cut the ball into ten slabs, we find:

$$L_{10} = 23.507611;$$
$$U_{10} = L_{10} + \frac{r\pi}{n} r^2 = 27.329987;$$
$$T_{10} = (L_{10} + U_{10})/2 = 25.418799.$$

The exact answer is 25.482505.

EXAMPLE: THE SHELL METHOD FOR A CONE

Instead of cutting our cone into slabs, we might have elected to cut it into cylindrical pieces. This means that we imagine the solid cone to be built up of concentric shells or tubes that telescope together to make up the whole cone (Figure 9.8). Then we estimate the volume of each shell and add them up.

Figure 9.8

To do this for our cone of radius r = 1.2 and height h = 3.4, we first rearrange it to have its base in the (x,y) plane and its axis along the z-axis. Then we subdivide the radius into 10 subintervals and consider the shells with inner radius $(i-1)r/10$ and outer radius $ir/10$, i = 1, 2, ..., 10. The height of the cone above the point x on the x-axis is $3.4 - \frac{3.4}{1.2} x$, so the shell lying between radii $1.2(i-1)/10$ and $1.2 \, i/10$ has a height between

Figure 9.9

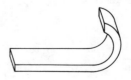

Figure 9.10

123

3.4(1 - (i-1)/10) and 3.4(1-i/10) (Figure 9.9). If we cut this
shell and lay it out flat, we see that its volume is approximately
its height times its thickness times its circumference (Figure 9.10).
Thus [3.4(11-i)/10][1.2/10][2π(1.2)(i-1)/10] is the approximation we
get from its inner height and circumference; [3.4(10-i)/10][0.12]
[0.24πi] is the outer estimate. (Notice that, in this special case
of a cone, these inner and outer estimates tend to be near each other,
since the height diminishes as the circumference increases.)

Let us calculate the sum S_{10} of these outer volumes for the
ten shells:

$$S_{10} = \frac{(3.4)(0.12)(0.24)\pi}{10} \sum_{i=1}^{10} i(10-i)$$

$$= 0.0307625 \sum_{i=1}^{10} i(10-i)$$

$$= 5.0758084.$$

The volumes corresponding to the inner measurements also add up to

$$0.0307625 \sum_{i=1}^{10} (i-1)(11-i) = S_{10}.$$

The integral corresponding to this shell method is

$$\int_0^{1.2} 3.4(1 - x/1.2)2\pi x \; dx = 6.8\pi \int_0^{1.2} (x - x^2/1.2) \; dx.$$

Its trapezoidal sum T_{10} is exactly the sum S_{10} that we have computed.

EXERCISES

1. Use the slab method to estimate volume in each of the following
cases. Divide the interval [0,5] of the x-axis into 5 subintervals,
each of length 1. For the ith slab, i = 1, 2, 3, 4, or 5, find the

124

area of the slice through the solid at $x = i$. Since each slab has unit thickness, the sum of these five areas is your volume estimate. In each case, compare your estimate with the theoretical volume (computed from the appropriate definite integral or otherwise).

*a. The cylinder: $0 \le x \le 5$ and $y^2 + z^2 \le 1$.

*b. The bar: $0 \le x \le 5$ and $0 \le y \le 1$ and $0 \le z \le 2$.

*c. The pyramid: $0 \le x \le 5$ and $|y| \le 5-x$ and $|z| \le 5-x$.

d. The thing: $0 \le x \le 5$ and $y^2 + z^2 \le 5 + (x-2.5)^2$.

2. A ball may also be thought of as made up of cylindrical shells. Estimate the volume of the same ball that we used in our Example of the slab method, with radius $r = 2.3$. Cut it into $n = 10$ shells and follow the Example of the shell method for the cone. The volume of each of its shells is approximately its height times its thickness times its circumference. The inner measurements for the tenth or

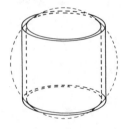

outside shell give

$$2 \sqrt{2.3^2 - 2.07^2} [0.23][4.14\pi],$$

for example. Find the approximate volumes of the other nine shells and add them up to get the sum S_{10} for the ball.

Figure 9.11

For what integral is S_{10} equal to the trapezoidal sum T_{10}?

*3. Use the slab method to find the volume of a solid whose base in the (x,y)-plane is the region between the graph of $y = \sin x$ and the interval $[0, \pi/2]$ of the x-axis, and for which each cross section or slice parallel to the (y,z)-plane is a quarter circle. First find the estimates of this volume corresponding to lower

Figure 9.12

and upper estimates for $n = 10$ slabs; then find an integral for
which these are just L_{10} and U_{10}. Use an antiderivative to evaluate
the integral and compare the theoretical volume with your trapezoidal
sum.

4. Calculate the volume of the solid swept out by rotating about
the x-axis the region below the graph of $y = x^2$ and above the interval [.12, .34]. Do this first by use of $n = 10$ slabs and then by
use of $n = 10$ shells. In each case, find the integral corresponding
to your sum and evaluate it theoretically as well.

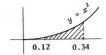

Figure 9.13

5. Revolve the region of Exercise 4 about the y-axis (instead of
the x-axis) to sweep out a solid. Calculate its volume by the use
of 10 slabs and also by use of 10 shells. In each case, find the
corresponding integral and evaluate it theoretically as well.

*6. Find the volume of the gold ring made by cutting a cylindrical
hole 11 mm in radius out of a sphere of radius 14 mm. Do this by
imagining that the region bet-
ween the graphs of $y = \sqrt{14^2 - x^2}$
and $y = 11$ is revolved about the
x-axis. Use 10 slabs first,
then find an integral corresponding to this sum and evaluate
it.

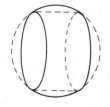

Figure 9.14

*7. Use 10 shells to find the volume of the solid swept out as the
triangle with vertices at $(0,0)$, $(1,0)$, $(2,1)$ is revolved about the
x-axis. Check your answer by calculating it as the difference in
volume of two cones.

126

*8. In our Example of the shell method the volume of a shell was estimated as height X thickness X circumference, and the trapezoidal sum averaged such volumes for inner and outer measurements. An alternative volume estimate is hA, where h is the average height and A is the exact cross-sectional area $A = \pi(r_{out}^2 - r_{in}^2)$, for outer and inner radii r_{out} and r_{in}. Recalculate the volume of the solid described in Exercise 7 using this new volume recipe for each shell. Which method is more accurate?

PROBLEMS

P1. Let a solid be generated by rotating about the y-axis the region bounded by $x = 0$, $y = 0$, $y = 1$, and the graph of $y = \ln(1/x)$. Find its volume using 10 slabs and then 10 shells.

P2. Estimate the volume of the ball of radius 2.3 as follows. Cut the half-ball described in our Example into 10 slabs. Then estimate the volume of each slab as the volume of a right circular cone cut off .23 units above its base. Then add up these estimates.

Figure 9.15

P3. Compute the modified trapezoidal sum C_{10} for the cone of our Example for the slab method, where $r = 1.2$ and $h = 3.4$, and compare your result with the theory. Attempt to do the same for the half-ball of radius 2.3.

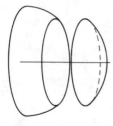

Figure 9.16

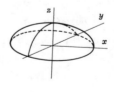

Figure 9.17

127

Next, reestimate the volume of the half-ball by making a preliminary slice at $x = 2.3/\sqrt{2}$. Hence you have two pieces obtained
by revolving about the x-axis the regions below the graph of
$y = \sqrt{2.3^2 - x^2}$ and above the intervals $[0,\ 2.3/\sqrt{2}]$ and $[2.3/\sqrt{2},\ 2.3]$.
Estimate the volume of the first piece by means of the modified
trapezoidal sum C_{10}. Next estimate the volume of the second piece,
the "spherical cap," by the slab method applied to the cap with its
cut face in the (x,y)-plane. Use 10 slabs and the modified trapezoidal sum C_{10}. Finally, add up your results and multiply by two
to approximate the volume of the whole ball. How accurate is your
answer?

P4. The *pressure* that a liquid exerts on a surface submerged in it
is proportional to the area of the surface and to the depth of the
liquid. Above one square centimeter of bottom area there lies 1 cc
of liquid per centimeter of depth. Since sea water weighs 1.025 g/cc,

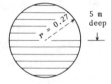

5 m
deep

Figure 9.18   Figure 9.19

the pressure at a depth of h cm is $1.025h$ g/cm^2 in the ocean. This
pressure is exerted in all directions, so that a round viewing port
at Marineland that is below water has an outward pressure on it corresponding to its depth. However, the surface of the port does not
all lie at any given depth.

To be specific, suppose that a port of radius $r = 0.27$ m has
its center 5 m below the surface. We approximate
the pressure on this port by subdividing the
vertical area into $n = 10$ narrow horizontal
strips, each of width 5.4 cm. The depth of the
bottom strip is 527 cm at its lower edge and
(527-5.4) cm along its upper edge. Its width

128

varies from 0 at the bottom to $2\sqrt{27^2 - 21.6^2}$ for its top edge. (We shall now suppress mention of the units of length or of weight for this problem; every measurement will be converted into centimeters or grams.) Accordingly, the pressure on this bottom strip may be estimated either as 0 or as $2\sqrt{27^2 - 21.6^2}$X5.4X521.6X1.025. More generally, the ith strip from the bottom of the port will have a pressure on it estimated at its top edge as $2\sqrt{27^2 - (27 - 5.4i)^2}$X 5.4X(527 - 5.4$i$)X1.025.

Add up these estimates for the ten strips to get the sum S_{10} estimating total pressure on the port. For what integral is this the trapezoidal sum T_{10}? How accurate is T_{10} as an estimate?

Answers to Starred Exercises

Exercises 1a. 15.707963, which is exact
           1b. 10, which is exact
           1c. 120 (166 2/3 is exact)
            3. $\pi/4 \int_0^{\pi/2} \sin^2 x \; dx = (\pi/4)^2 = 0.6168503$
            6. $2\pi \int_0^{\sqrt{75}} (75-x^2) dx = 2720.6991$ mm^3 = 2.7206991 cc
            7. $U_{10} = L_{10} = 1.0367256$ ($\pi/3$ is exact; error is 1%)
            8. $T_{10} = 1.0524335$ ($\pi/3$ is exact, error is ½%)

10

CURVES AND POLAR COORDINATES

INTRODUCTION

As we mentioned in the Introduction to Chapter 9, the calculation of lengths of curved lines was one of the principal problems that led to the creation of the calculus. It was an old and intractable problem. Archimedes had used polygons inscribed in a circle to calculate π, but nothing further was discovered about curve lengths until the seventeenth century. In fact, even such a powerful mathematician as Descartes (1596-1650) had asserted that the length of no curve but the circle would ever be calculated. He was proven wrong, however, first by Torricelli in his work on the logarithmic spiral and then by the English architect, Christopher Wren, who established the length of the cycloid.

These and other particular results provided some of the setting in which Sir Isaac Newton (1642-1727) and Baron Gottfried Wilhelm von Leibniz (1646-1716) worked. In fact, Newton said that if he saw further than other men, it was only because he stood on the shoulders of giants. We shall see Examples of what he saw, first for the two functions $f(x) = 2\sqrt{x}$ and $g(x) = x^2/4$. Then the exponential spiral provides our Example of parametric equations for a curve while Archimedes' spiral illustrates the calculation of curve lengths in polar coordinates.

In the Exercises we will explore further calculations of length for these same curves as well as for the parabola and the cycloid. In the Problems we will examine how the cardioid and the ellipse are measured for length.

EXAMPLE: $f(x) = 2\sqrt{x}$

How long is the curved line that is the graph of the function $f(x) = 2\sqrt{x}$, say between $(0,0)$ and $(1,2)$? That is, these two points are on that graph (see Figure 10.1), and the distance between $(0,0)$ and $(1,2)$ is $\sqrt{(1-0)^2+(2-0)^2} = \sqrt{5}$. Thus $\sqrt{5}$ is the length of the dotted line in Figure 10.2; the curved line is certainly longer, but how long is it? If Figure 10.2 represented a map of a curved road between two towns, the dotted line would represent the distance between the towns "as the crow flies." The problem is to find the distance along the road.

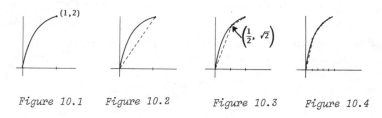

Figure 10.1   Figure 10.2   Figure 10.3   Figure 10.4

There was a similar question at the beginning of Chapter 7, where we asked ourselves how to measure π, the distance along the curve of a semicircle. We avoided this question then in favor of a theoretical argument that led us to compute the area of a disc in order to measure π. However, we now address the question squarely: how do we find the length of a curve?

Suppose we choose a point midway, say $x = \frac{1}{2}$ and $y = 2\sqrt{\frac{1}{2}} = \sqrt{2}$, and calculate the lengths of the two dotted straight lines of

Figure 10.3: $\sqrt{\tfrac{1}{2}^2 + \sqrt{2}^2} + \sqrt{\tfrac{1}{2}^2 + (2-\sqrt{2})^2} = 2.2701596$. This is, of course, a larger number than $\sqrt{5} = 2.2360680$, which is the length of the single straight line from $(0,0)$ to $(1,2)$. But the curve is longer than the sum of the lengths of the two dotted lines in Figure 10.3, as well. If we divide the interval $[0,1]$ of the x-axis into five sub-intervals of equal length, we can define the broken dotted line of Figure 10.4 that approximates the curve with five short chords. In order to calculate the distance along this dotted route, we organize our computations with the sigma notation. The total length is:

$$\sum_{i=1}^{5} \left[(x_i - x_{i-1})^2 + (2\sqrt{x_i} - 2\sqrt{x_{i-1}})^2 \right]^{\frac{1}{2}}$$

$$= \sum_{i=1}^{5} \left[1 + \left(\frac{2\sqrt{x_i} - 2\sqrt{x_{i-1}}}{x_i - x_{i-1}} \right)^2 \right]^{\frac{1}{2}} (x_i - x_{i-1}).$$

We calculate this number and record it in Table 10.1 with our previous results for a single chord (as in Figure 10.2) and two chords (as in Figure 10.3). We list also the similar calculation for ten chords, which is fairly hard work.

TABLE 10.1

Number of Chords	Total Length of Chords
1	2.2360680
2	2.2701596
5	2.2881026
10	2.2927511

In order to understand the limit process underlying our calculations, let us inspect the sum for the lengths of the chords over n equal subintervals:

132

$$\sum_{i=1}^{n} \left[1 + \left(\frac{f(x_i) - f(x_{i-1})}{x_i - x_{i-1}} \right)^2 \right]^{\frac{1}{2}} (x_i - x_{i-1}).$$

The mean value theorem says that there is a number ξ_i between x_{i-1} and x_i such that $f(x_i) - f(x_{i-1}) = (x_i - x_{i-1}) f'(\xi_i)$; consequently, our sum may be restated as

$$\sum_{i=1}^{n} \left[1 + f'(\xi_i)^2 \right]^{\frac{1}{2}} (x_i - x_{i-1}).$$

In this form, this sum is seen to be a Riemann sum for the integral $\int_0^1 \left[1 + f'(x)^2 \right]^{\frac{1}{2}} dx$. Therefore the limit as n tends to infinity of the sum of the lengths of the approximating chords is this integral.

The integral may be rewritten as $\int_0^1 (1+1/x)^{\frac{1}{2}} dx$. If the substitution $x = \tan^2 \theta$ is made, the resulting expression $\int_0^{\pi/4} 2\sec^3\theta d\theta$ may be integrated by parts. This theoretical length, the limit of the lengths of chordal approximations to $f(x) = 2\sqrt{x}$ over $[0,1]$, is $\sqrt{2} + \ln(\sqrt{2} + 1) = 2.2955871$.

We remark that it is typical of curve length problems that the antiderivative of $[1+f'(x)]^{\frac{1}{2}}$ is difficult to find, even for an elementary function f. Numerical methods thus often offer our only hope. With this in mind, we have calculated the sum, as in Table 10.1 but for 20 chords, to be 2.2945372; for 100 subintervals it is 2.2954877. Do not attempt these calculations yourself (unless your machine is programmable); we cite them to show that a hundred-chord approximation has only 0.004% error in this case.

EXAMPLE: $g(x) = x^2/4$

Suppose we interchange the roles of x and y in our curve to get the function $g(x) = x^2/4$ on the interval $[0,2]$. The graph of g is merely the graph of f reflected in the diagonal line $y = x$ (see Figure 10.5). Thus the length of this segment of the graph of g is

exactly the length we have been investigating (g is the inverse
function for f). And, of course, the distance in Figure 10.6 from
end to end of this segment of the graph of g is $\sqrt{5}$, just as before.

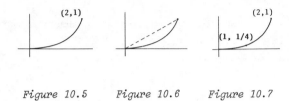

Figure 10.5   *Figure 10.6*   *Figure 10.7*

But the distance along the chords in Figure 10.7 is

$$\sqrt{1^2 + \tfrac{1}{4}^2} + \sqrt{1^2 + (1-\tfrac{1}{4})^2} = 2.2807764;$$

this is not the same estimate as the case $n = 2$ for f. This difference tells us that our subdivisions of the *curve* are not of equal
length along the curve or along the chords. The *equal* intervals are
along one or the other axis. Exercise 2 asks for the estimates for
g in case $n = 5$ or 10; clearly the limiting value for the length of
g over [0,2] will be exactly the limiting value for f over [0,1].
Nevertheless, along the way, the finite stages for f and for g will
differ.

EXAMPLE: PARAMETRIC EQUATIONS AND THE EXPONENTIAL SPIRAL

Suppose we describe the movement of a particle in the plane by giving
its x and y coordinates at various times t, say $x(t) = e^t \cos t$ and
$y(t) = e^t \sin t$. Then the path of the particle is not the graph of
the function $x(t)$, nor of the function $y(t)$, but rather a simultaneous
graph for both equations. You can picture this curve by remembering
that the point with coordinates ($\cos t$, $\sin t$) lies on the circle of
radius 1 centered at the origin. Thus ($\cos t$, $\sin t$) corresponds to
the point t radians counterclockwise from the x-axis. Hence
($e^t \cos t$, $e^t \sin t$) lies on the same line from the origin but at a
distance e^t along that line. Since $0 \leqq t$ means $1 \leqq e^t$, as t increases

134

from 0 the particle spirals outward from (1,0), going counterclockwise around the origin as it goes away from the origin. Figure 10.8
shows this curve, called the *exponential spiral*, for $0 \leq t \leq \ln 2$.
The distance along a chord from end to end of this curve segment is

$$[(2 \cos \ln 2 - 1)^2 + (2 \sin \ln 2 - 0)^2]^{\frac{1}{2}} = 1.3867388.$$

To make a better estimate of the distance the particle travels
along the curve between time 0 and time $t = \ln 2$, we choose the midpoint *in time*, $t = \ln(2)/2 = \ln\sqrt{2}$, and find a point (1.33, 0.48)
along the curve with which to make the dotted line approximation of

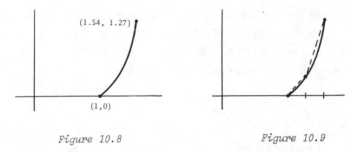

Figure 10.8   *Figure 10.9*

Figure 10.9. The sum of the lengths of these two dotted chords is

$$[(\sqrt{2} \cos \ln\sqrt{2} - 1)^2 + (\sqrt{2} \sin \ln\sqrt{2})^2]^{\frac{1}{2}}$$
$$+ [(2 \cos \ln 2 - \sqrt{2} \cos \ln\sqrt{2})^2 + (2 \sin \ln 2 - \sqrt{2} \sin \ln\sqrt{2})^2]^{\frac{1}{2}}$$
$$= 1.4071887.$$

For more subintervals we utilize the sigma notation to condense
unwieldy expressions. If the time interval [0, ln 2] is divided into
n equal pieces, the sum of the lengths of chords is

$$\sum_{i=1}^{n} \left(\left[x(t_i) - x(t_{i-1}) \right]^2 + \left[y(t_i) - y(t_{i-1}) \right]^2 \right)^{\frac{1}{2}}.$$

135

We can compute this sum for n = 5. The results are presented in
Table 10.2 along with our previous sums.

TABLE 10.2

Number of Chords	Total Length of Chords
1	1.3867388
2	1.4071887
5	1.4130825

We note that the sum for n = 5 was a challenge to do properly; our
first two attempts failed (absurd answers resulted from some incorrect keying operation). In order to evaluate the limit of these sums
as $n \to \infty$, we both multiply and divide by the increment in t to display the sum as

$$\sum_{i=1}^{n} \left[\left(\frac{x(t_i) - x(t_{i-1})}{t_i - t_{i-1}} \right)^2 + \left(\frac{y(t_i) - y(t_{i-1})}{t_i - t_{i-1}} \right)^2 \right]^{\frac{1}{2}} (t_i - t_{i-1}).$$

Notice here that

$$\frac{x(t_i) - x(t_{i-1})}{t_i - t_{i-1}}$$

is a difference quotient that approaches $x'(t_i)$, the derivative of
the function $x(t)$ at $t = t_i$, as $t_{i-1} \to t_i$. Similarly, the summand

$$\frac{y(t_i) - y(t_{i-1})}{t_i - t_{i-1}}$$

has limit $y'(t_i)$; it can be shown that the sum has for its limit as
$n \to \infty$ the integral

136

$$\int_0^{\ln 2} [x'(t)^2 + y'(t)^2]^{\frac{1}{2}} \, dt.$$

POLAR COORDINATES

Let (x,y) be the ordinary *Cartesian* or *rectangular coordinates* of a point in the plane: the *polar coordinates* of that same point are a number pair (r,θ) such that $x = r \cos \theta$ and $y = r \sin \theta$. There are other polar coordinates for that same point; $(-r,\ \theta+\pi)$ is another pair that works as well as (r,θ), and $(r,\ \theta+2\pi)$ is still another. If f is a real valued function of real numbers, the *polar graph* of f is the graph whose parametric equations are $\theta = t$ and $r = f(t)$, where (r,θ) are the polar coordinates of the point at "time" t. We could simplify this statement by suggesting that "θ be taken as parameter" to graph $(f(\theta),\ \theta)$ in polar coordinates.

For instance, if $f(x) = e^x$, then the polar graph of f is the set of points having polar coordinates (e^t, t). The rectangular coordinates for this graph are $x = e^t \cos t$ and $y = e^t \sin t$! (Be sure you understand this.) This is the parametric curve of our previous example (see Figure 10.8). Another example is the polar graph of the function $f(x) = x$: parameterized by θ this is a graph of the equation $r = \theta$. This is called the *spiral of Archimedes*; it is shown

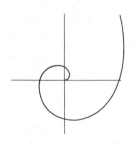

Figure 10.10

in Figure 10.10. To view this curve from a different perspective, we translate its parametric equation into parametric equations for the corresponding rectangular coordinates. Since the general rule is $x = r \cos \theta$ and $y = r \sin \theta$, the parametric curve is $(x,y) = (t \cos t,\ t \sin t)$; here we have replaced the symbol θ by the symbol t.

Conversely, if we translate the rectangular graph of the identity function $f(x) = x$ into polar coordinates, we use the general rule

$r = \sqrt{x^2+y^2}$ and $\theta = \arctan\left(\dfrac{y}{x}\right)$. This gives $r(t) = \sqrt{2}\, t$ and $\theta(t) = \arctan 1 = \dfrac{\pi}{4}$. Here we have replaced the parameter x by the parameter t; the graph in polar coordinates is the straight line through the origin with slope 1. It is the same graph as it was when it was *described* in rectangular coordinates, and it is *not* the spiral of Archimedes.

Now suppose we would like to find the length of a segment of the polar graph of a function. One solution is to translate the polar equation into two rectangular ones. Sometimes that's the easiest way to solve this problem. However, it may well be easier to attack the polar coordinate problem directly as in Figure 10.11. Here we wish

Figure 10.11

to measure the distance along a chord from the point with polar coordinates (r_0,θ_0) to (r_1,θ_1). The length of the arc from (r_0,θ_0) to (r_0,θ_1) is $r_0(\theta_1-\theta_0)$, and for small enough increments $\theta_1 - \theta_0$ in the parameter this number is a good estimate (see Exercise 4) for the length of the straight line from (r_0,θ_0) to (r_0,θ_1). The distance from (r_0,θ_1) to (r_1,θ_1) is of course $r_1 - r_0$. The arc from (r_0,θ_0) to (r_0,θ_1) meets the radius $\theta = \theta_1$ at right angles. Hence the Pythagorean theorem says that the length of the chord is approximately $\left[r_0{}^2(\theta_1-\theta_0)^2 + (r_1-r_0)^2\right]^{\frac{1}{2}}$.

Example: The Spiral of Archimedes

Let us apply this recipe to calculate the length of the spiral of Archimedes between $\pi/2$ and π. The two ends of this curve segment have rectangular coordinates $(0, \pi/2)$ and $(\pi,0)$ so the distance between them is $[(\pi/2)^2+\pi^2]^{\frac{1}{2}} = 3.5124074$ (Figure 10.12). The recipe gives 2.9249734; this is a very rough approximation to the length of the chord itself. Nevertheless, as the points get closer together (as in Figure 10.13), the recipe should give a better approximation.

138

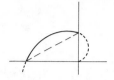 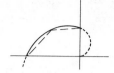

Figure 10.12   *Figure 10.13*

We wish to calculate the lengths of chordal approximations given by
our recipe, which for n chords is

$$\sum_{i=1}^{n} \left[r_{i-1}^2 (\theta_i - \theta_{i-1})^2 + (r_i - r_{i-1})^2 \right]^{\frac{1}{2}}.$$

In the case of the spiral, $r = \theta$ and the subintervals are of
length $\theta_i - \theta_{i-1} = (\pi - \pi/2)1/n = \pi/2n$. Hence

$$r_{i-1} = \frac{\pi}{2} + (i-1)\frac{\pi}{2n} = \frac{(n+i-1)\pi}{2n}$$

and

$$r_i - r_{i-1} = \pi/2n.$$

Our recipe thus becomes (here we have adjusted the range of the
summation index downward by 1, for clarity):

$$\sum_{i=0}^{n-1} \left[\left(\frac{(n+i)\pi}{2n}\right)^2 \left(\frac{\pi}{2n}\right)^2 + \left(\frac{\pi}{2n}\right)^2 \right]^{\frac{1}{2}}$$

$$= \frac{\pi}{2n} \sum_{i=0}^{n-1} \left[1 + \left(\frac{(n+i)\pi}{2n}\right)^2 \right]^{\frac{1}{2}}$$

We have tabulated our results in Table 10.3.

TABLE 10.3

n	Total Length
1	2.9249734
2	3.4728072
5	3.8062480
10	3.9182658
20	3.9744423

In order to compute a theoretical result, we divide each summand of the first recipe above by $\theta_i - \theta_{i-1}$ to get

$$\sum_{i=1}^{n} \left[r_{i-1}{}^2 + \left(\frac{r_i - r_{i-1}}{\theta_i - \theta_{i-1}} \right)^2 \right]^{\frac{1}{2}} \left(\theta_i - \theta_{i-1} \right).$$

Again the MVT tells us that there is a number ξ_i in the ith subinterval such that

$$\frac{r(\theta_i) - r(\theta_{i-1})}{\theta_i - \theta_{i-1}} = r'(\xi_i).$$

Therefore this sum is a Riemann sum for the integral

$$\int_a^b [r(\theta)^2 + r'(\theta)^2]^{\frac{1}{2}} \, d\theta.$$

In our case, where $r(\theta) = \theta$ and $r'(\theta) = 1$, we have $\int_{\pi/2}^{\pi} (\theta^2+1)^{\frac{1}{2}} \, d\theta$.

Exercise 5 asks that you show that this gives a theoretical length of 4.0307311.

Compare this theoretical, precise length with our calculated estimate for $n = 20$: the estimate is about $1\frac{1}{2}\%$ low. Since it is not difficult to find an antiderivative for $(\theta^2+1)^{\frac{1}{2}}$, the ease with

140

which the theory provides precise answers is impressive. On the
other hand, the recipe will provide us with good estimates, even
when no antiderivative is in sight.

EXERCISES

1. For each function $f(x)$, estimate the length of the graph over
[0,1] by approximations with 1, 2, and 5 chords. Then find the exact
length by use of the integral formula.

  *a. $f(x) = x$ *c. $f(x) = -\ln \cos x$

  *b. $f(x) = x^{3/2}$ d. $f(x) = \frac{1}{2}(e^x + e^{-x})$

*2. Find the lengths of the chordal approximations for the function
$g(x) = x^2/4$ (of our Example above) over [0,2] when 5 and then 10
chords are used. Finally, integrate $\int_0^2 \sqrt{1+g'(x)^2}\ dx$ and evaluate the
limiting length of this curve segment.

3. Duplicate the calculation of our Example of the length of the
parametric curve $x(t) = e^t \cos t$, $y(t) = e^t \sin t$. That is, find the
lengths of the approximations by 1, 2, and 5 chords. Then use an
antiderivative to evaluate the appropriate integral on [0, ln 2] and
compare this theoretical result to your calculated estimates.

4. Show that in Figure 10.11 the distance from (r_0,θ_0) to (r_0,θ_1) is
exactly $2\ r_0 \sin \left(\frac{\theta_1-\theta_0}{2}\right)$. Then prove that as $\theta_1 \to \theta_0$, this exact
length approaches our estimate $r_0(\theta_1-\theta_0)$. Do this by showing that

$$\lim_{\varphi \to 0} \frac{2\ \sin(\varphi/2)}{\varphi} = 1.$$

5. Verify the calculations of the Example to approximate the length
of Archimedes' spiral between $\pi/2$ and π. Then find an antiderivative
for $\sqrt{\theta^2+1}$ and so calculate the theoretical length to be 4.0307311.

*6. Compute the length of the approximation by five chords to the
spiral of Archimedes between 2π and $5\pi/2$. Then compare your estimate to the limiting length given by the integral.

7. If a ball is thrown horizontally from the top of a tall building, Newton's laws of motion will describe its travel. If we choose a coordinate system with origin at the building's top, then its horizontal coordinate will change at a constant rate, which is the velocity imparted by the throw. Its vertical coordinate at time t will be negative as the ball falls, and its size will be proportional

to t^2. Suppose a given throw results in the ball taking the path described parametrically in feet at time t by $x(t) = 64t$ and $y(t) = -16t^2$. Find the time at which the ball hits ground if the building is 100 feet high. Estimate the total length of its path by use of approximations by $n = 1, 2, 5,$ and 10 chords. Then find an anti-derivative for the appropriate integral and so get an exact answer. Compare this answer to your estimates.

*8. The path a nail in a tire travels as the tire rolls is called a *cycloid*. We imagine that the center of the tire (the center of the axle) is moving at a constant speed in order to parameterize this curve. Since the tire doesn't slip on the road, when the bottom of the tire is at a point t on the x-axis, the angle tOP is equal to t. Show that the x-coordinate of P is

Figure 10.14

$$x(t) = t - \sin t$$

and that the y-coordinate of P is $y(t) = 1 - \cos t$.

Now estimate how far the nail travels when the car goes 1 mile. That is, assume the tire has radius 1 foot, find the length L of one arch of the cycloid and then compute $L/2\pi$ = the ratio of the length of the arch of the cycloid to the base length. Do this by estimating $L/2$ over the interval $[0,\pi]$ with $n = 1, 2,$ and 5 chords. Then find an antiderivative for the

142

appropriate integrand and so compute the exact value of $L/2\pi$.

PROBLEMS

*P1.   Consider $f(x) = 1 + \cos x$:   the "rectangular" graph of f is the wavy line shown in Figure 10.15.   The polar graph of f is called a *cardioid*.   The complete graph, shown in Figure 10.16, is tracked out by $(f(\theta), \theta)$ as θ goes from 0 to 2π.   Find the length of the segment of the cardioid in Figure 10.16 that lies in the first quadrant.

Figures 10.15 and 10.16

That is, find the length of the polar graph $(1 + \cos \theta, \theta)$ for $0 \leq \theta \leq \pi/2$.   Do this by subdividing the domain $[0, \pi/2]$ of the parameter θ into $n = 2$, then 5, then 10 equal parts and estimating curve length to be

$$\sum_{i=1}^{n} \{(1 + \cos[(i-1)\varphi])^2 \varphi^2 + [\cos i\varphi - \cos(i-1)\varphi]^2\}^{\frac{1}{2}}$$

where φ stands for $\pi/2n$.   Then find the limiting theoretical value for this curve length by integration and compare your estimates with it.

*P2.   The parametric equations for an ellipse are $x = a \cos t$ and $y = b \sin t$ for $0 < b \leq a$ and $0 \leq t \leq 2\pi$.   Show that the total length of an ellipse is

143

$$4a \int_0^{\pi/2} (1 - e^2 \cos^2 t)^{\frac{1}{2}} dt$$

where e is the *eccentricity* of the ellipse, $e = \sqrt{a^2-b^2}/a < 1$. This is called an *elliptic integral (of the second kind)*; it has no elementary antiderivative. Estimate the integral in order to estimate the length of the parametric ellipse for $0 \leq t \leq \pi/2$ when $a = 1$ and $e = \frac{1}{2}$; use the trapezoidal sums T_2 and T_5.

Observe the small error of your results in relation to the correct answer, which is 5.8698488. Can you explain this by referring to the modified sums C_n that are discussed in Problem P1, Chapter 7?

Answers to Starred Exercises and Problems

Exercises 1a. All lengths are $\sqrt{2} = 1.4142136$.

1b. 1.4142136 for $n = 1$, 1.4296198 for $n = 2$, 1.4372275 for $n = 5$; 1.4397099 for the integral

1c. 1.1743066 for $n = 1$, 1.2133813 for $n = 2$, 1.2241467 for $n = 5$; 1.2261912 for the integral

2. 2.2932281 for $n = 5$, 2.2949978 for $n = 10$; 2.2955871 for the integral

6. 10.970071 for $n = 5$; 11.214313 for the integral

8. 1.1854471 for $n = 1$, 1.2445528 for $n = 2$, 1.2681480 for $n = 5$; $4/\pi = 1.2732395$ for the integral

Problems P1. 3.1136643 for $n = 2$, 2.9550987 for $n = 5$, 2.8939200 for $n = 10$; 2.8284271 for the integral

P2. $T_2 = 5.8698367$, $T_5 = 5.8698488$

144

11

SEQUENCES AND SERIES

INTRODUCTION

Sequences and series have fascinated people for thousands of years.
They are arrows pointing at the unreachable infinite. Aristotle
described the paradoxes due to Zeno, of Achilles racing the tortoise
and of "dichotomy," both of which are answerable today as questions
about infinite series. And Archimedes understood that the geometric
series $1 + \frac{1}{4} + \frac{1}{4^2} + \frac{1}{4^3} + \ldots$ was the number 4/3. But there was very
little more than that known, in theory or practice, to guide Isaac
Newton when he went to work on the calculus. He used series in
wholly new ways, applying his techniques of integration and differentiation to them term by term.

Newton was the first to derive the series expansions for many
elementary functions. In fact, series were often the only way he
could deal with these functions theoretically. For some time, series
also offered the best computational methods for the values of the log
and trig functions.

Series are a very powerful method for use with a calculator. They are lots of fun to sum, too, so they are a pleasant way to learn much about the calculus. We shall begin with the definition used today, although modern notions of series convergence were not available to the men who discovered these facts we shall study. This definition is immediately applied to study the harmonic and p-series.

Next we shall see Examples of geometric series and of alternating series. Further Examples illustrate the estimation of remainders and the acceleration of convergence using several different techniques. The Exercises give practice in forming partial sums and estimating remainders. One Exercise investigates the ratios of successive terms of the Fibonacci sequence.

In the Problems we will develop some theory for the study of convergence. We will also investigate a Fourier sin series, Stirling's formula, the Euler number, and continued fractions. Finally two Problems provide a closed formula and a generating function for the Fibonacci sequence.

THE DEFINITIONS

A *series* or "infinite sum" of numbers $a_1 + a_2 + a_3 + \ldots + a_n + \ldots = \sum_{i=1}^{\infty} a_i$ is, by definition, merely a sequence: namely, the sequence $a_1, a_1 + a_2, a_1 + a_2 + a_3, \ldots, a_1 + a_2 + \ldots + a_n, \ldots$. Each *partial sum* $S_n = a_1 + a_2 + \ldots + a_n$ may be expressed in sigma notation as $S_n = \sum_{i=1}^{n} a_i$. It is the sum of the first n *terms* of the series. We will usually denote the series $\sum_{i=1}^{\infty} a_i$ simply by $\sum_1 a_i$. It *converges* to S or has the *sum* S if the associated sequence of partial sums has limit S; otherwise the series is said to *diverge*.

EXAMPLE: THE HARMONIC SERIES

Of course, we have already seen that our calculator cannot tell us whether a sequence converges or not. An example of a divergent sequence that doesn't look that way to our machine is the familiar *harmonic series,*

$$\sum_1 \frac{1}{i} = 1 + \frac{1}{2} + \frac{1}{3} + \frac{1}{4} + \ldots .$$

We have calculated S_{10} to be 2.9289683, yet S_{100} is only 5.1774272.
Since the next hundred terms have a sum less than 1, the third hundred a sum less than ½, the fourth hundred a sum less than 1/3, and
so on (do you understand why?), we see that $S_{1100} < S_{100} + S_{10} =$
8.1063955. You might think that if the first 1100 terms add up to
less than 9 and each subsequent term is less than 1/1100, surely
there would be a limit. Nevertheless, we can easily prove that these
partial sums S_n, for large enough index n, become larger than 1000,
larger than 1,000,000; and eventually larger than any preassigned
number. This is true because the second through tenth terms are all
at least as large as their last one, 1/10. Hence $S_{10} > 1 + 9/10$.
The 90 terms between 11 and 100 add up to at least $90/100 = 9/10$,
$\sum_{i=101}^{1000} a_i > 900/1000 = 9/10$, and so on, to give $S_{10^n} > 1 + (9/10)n$.
Obviously the sequence $1 + (9/10)n$ diverges, so $\sum_1 1/i$ diverges.

Example: p-Series

It is a THEOREM that *the p-series*

$$\sum_1 \frac{1}{i^p} = 1 + \frac{1}{2^p} + \frac{1}{3^p} + \ldots ,$$

where the exponent p *is a fixed number, converges if and only if*
$p > 1$. Hence $\sum_1 1/i^2$ converges; in fact Leonhard Euler showed in
1734 that its sum is $\pi^2/6$. Nevertheless, the twentieth partial sum
here is only $S_{20} = 1.5961632$, yet $\pi^2/6 = 1.6449341$. This is an error
of 3%; it is clear that Euler did not guess this sum by adding up a
few terms. (Actually he derived it from a product expansion for
sin x, plus years of thought on the matter.) A more satisfying

p-series for calculating π was also given by Euler:

$$\sum_1 1/i^4 = \pi^4/90.$$

We list some partial sums in Table 11.1.

TABLE 11.1

n	S_n
5	1.0803519
10	1.0820366
15	1.0822339
20	1.0822846

$$\vdots$$

$$\pi^4/90 \ = \ 1.0823232$$

GEOMETRIC SERIES

It is algebraically simple to find an exact sum for the *geometric series* $\sum_0 ar^i$ for any numbers a and r (here we introduce the notation $\sum_0 ar^i$ for $\sum_{i=0}^{\infty} ar^i$).

Remember that

$$1 - r^{n+1} = (1-r)(1 + r + r^2 + \ldots + r^n)$$

(or just multiply it out to check it), so $1 + r + r^2 + \ldots + r^n = \dfrac{1 - r^{n+1}}{1 - r}$ whenever $r \neq 1$. (We have seen this sum once before in Problem P2, Chapter 8 in which we calculated the monthly payments due on a loan.) Hence the partial sum S_n for the geometric series is $S_n = \dfrac{a - ar^{n+1}}{1 - r}$. A moment's reflection shows that if $|r| \geq 1$, the sequence of partial sums had no limit; the series converges to $a/(1-r)$ if and only if $|r| < 1$.

148

Example: An Alternating Series

It is often easy to decide about convergence for *alternating series*, which are series whose terms alternate in sign. They converge if the absolute values of the terms themselves form a decreasing sequence with limit 0. In that case, the error $|S_n - S|$ between the nth partial sum and the limit is less than the absolute value of the next term $|a_{n+1}|$.

An example is

$$\sum_1 (-1)^{i+1} \frac{2i+1}{i^2+i} .$$

Clearly

$$\lim_{n\to\infty} \frac{2n+1}{n^2+n} = \lim_{n\to\infty} \frac{2/n + 1/n^2}{1 + 1/n} = 0,$$

so the series is convergent. If we examine a few partial sums (see Table 11.2), we form a suspicion that the limit is 1. Encouraged by

TABLE 11.2

n	S_n
5	1.1666667
10	0.9090909
15	1.0625000
20	0.9523810

this to think a little, we theoretically investigate the nth term,

$$(-1)^{n+1} \frac{2n+1}{n^2+n} = (-1)^{n+1} \left[\frac{2}{n+1} + \frac{1}{n^2+n} \right]$$

$$= (-1)^{n+1} \left[\frac{2}{n+1} + \frac{1}{n} - \frac{1}{n+1} \right]$$

$$= (-1)^{n+1} \left[\frac{1}{n} + \frac{1}{n+1} \right].$$

But then any one of its partial sums may be rewritten with two terms for each index to give

$$S_n = (1 + 1/2) - (1/2 + 1/3) + (1/3 + 1/4) - \ldots +$$

$$(-1)^{n+1}(1/n + 1/(n+1)) = 1 + (-1)^{n+1} 1/(n+1).$$

Clearly $S_n \to 1$.

EXAMPLE: ESTIMATION OF REMAINDERS BY INTEGRALS

The difference $S - S_n = R_n$ between a series and its nth partial sum is called a *remainder* or *truncation error*. Since $S = S_n + R_n$, R_n itself is an infinite series, whose exact determination is conceptually as difficult as the evaluation of S itself. Nevertheless, a crude estimate for R_n can result in a sharper estimate than S_n for S. To see this, reconsider the slowly convergent p-series for $p = 2$,

$\sum_1 1/i^2 = \pi^2/6$. Our calculations above showed that S_{20} was in error

by 3%.

We may estimate R_n by *comparison with an integral*. For reasons similar to those that establish the integral test for convergence, the improper integral $\int_n^\infty 1/x^2 \, dx$ is greater than R_n (Figure 11.1), and also $\int_{n+1}^\infty 1/x^2 \, dx < R_n$ (Figure 11.2).

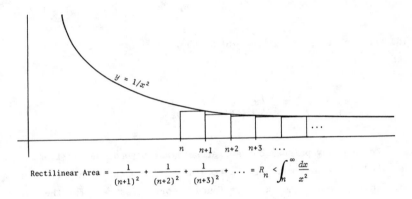

Rectilinear Area $= \dfrac{1}{(n+1)^2} + \dfrac{1}{(n+2)^2} + \dfrac{1}{(n+3)^2} + \ldots = R_n < \displaystyle\int_n^\infty \frac{dx}{x^2}$

Figure 11.1

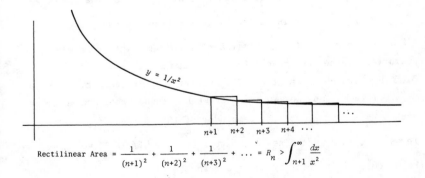

Rectilinear Area $= \dfrac{1}{(n+1)^2} + \dfrac{1}{(n+2)^2} + \dfrac{1}{(n+3)^2} + \ldots = R_n > \displaystyle\int_{n+1}^\infty \frac{dx}{x^2}$

Figure 11.2

151

Now

$$\int_n^\infty 1/x^2 \, dx = \lim_{N\to\infty} \int_n^N 1/x^2 \, dx = \lim_{N\to\infty}\left(\frac{1}{n} - \frac{1}{N}\right) = \frac{1}{n}$$

evaluates this integral. Hence we have $1/(n+1) < R_n < 1/n$ as upper and lower bounds for R_n. Suppose that we estimate R_n to be the average,

$$\frac{1}{2}\left(\frac{1}{n+1} + \frac{1}{n}\right) = \frac{2n+1}{2n^2+2n} \ .$$

The error of this estimate could be no larger than half the difference,

$$\frac{1}{2}\left(\frac{1}{n} - \frac{1}{n+1}\right) = \frac{1}{2n^2+2n} \ .$$

We have now pursued our answer somewhat beyond the question. We asked for an estimate of the remainder R_n, which is the error in our finite sum S_n. But we have found a correction term for the partial sum. That is, for the p-series $\sum_1 1/i^2$, if we add $(2n+1)/(2n^2+2n)$ to S_n to get a new estimate T_n for S, the error $|S-T_n| < 1/(2n^2+2n)$. In particular, if $n = 20$, we have $T_n = 1.6449727$, compared to the limit $S = 1.6449341$. The error was guaranteed to be less than $1/840 = 0.0011905$; it is in fact 0.0000386, which is about 0.002%, a startling improvement.

Our technique thus worked once we had found upper and lower bounds $J_n \leq R_n \leq K_n$ for the remainder of this series of positive terms. We then made a new estimate $T_n = S_n + \frac{1}{2}(J_n+K_n)$, which we knew would be in error by no more than $\frac{1}{2}(K_n-J_n)$.

EXAMPLE: ESTIMATION OF REMAINDERS FOR ALTERNATING SERIES

If we substitute $-r$ for r in the geometric series, we see that

$$1/(1+r) = 1 - r + r^2 - \ldots + (-1)^{n-1}r^{n-1} + (-1)^n r^n/(1+r) \quad \text{if } r \neq -1$$

(the last term is the remainder). Integrating both sides of this equality from 0 to some number $x > 0$ we find that

$$\ln(1+x) = x - \frac{x^2}{2} + \frac{x^3}{3} - \ldots + (-1)^{n-1}\frac{x^n}{n} + R_n$$

where

$$R_n = (-1)^n \int_0^x \frac{r^n dr}{1+r}$$

is the nth remainder term for a new series $\displaystyle\sum_1 (-1)^{i-1}\frac{x^i}{i}$.

Since $x > r > 0$ inside the interval of integration we can estimate $|R_n| = (-1)^n R_n$ by

$$\int_0^x \frac{r^n}{1+x}\, dr < (-1)^n R_n < \int_0^x r^n\, dr$$

or

$$\frac{x^{n+1}}{(1+x)(n+1)} < (-1)^n R_n < \frac{x^{n+1}}{n+1} .$$

Clearly $\displaystyle\lim_{n\to\infty} R_n = 0$ if and only if $x \leq 1$, and this is exactly the condition for the convergence of the series to $\ln(1+x)$. (Since the series is alternating, it converges because its terms decrease in size and have limit zero. Then $|R_n|$ is automatically less than the next term $x^{n+1}/(n+1)$.)

It is important to be aware that this series is like the geometric series in that the terms of the series are themselves functions that depend on a variable x. Each term here is of the general form $a_i x^i$, where a_i is a number depending only on i; a series of such terms is called a *power series*.

But we may again use our estimates for R_n to sharpen our partial sum estimates for S itself. This time the series is alternating, and R_n alternates in sign. Since our *lower* bound $\frac{x^{n+1}}{(1+x)(n+1)} < |R_n|$, if we add a new term to S_n which is this lower bound with signature opposite to that of the nth summand, the result

$$T_n = S_n + (-1)^n \frac{x^{n+1}}{(1+x)(n+1)}$$

will lie between S_n and S. Thus the sequence $\{T_n\}$ alternates about S just as does $\{S_n\}$, and its error is

$$|S-T_n| \leq |T_n - T_{n+1}| = \frac{x^{n+2}}{(1+x)(n+1)(n+2)} < \frac{1}{n^2} \frac{x^{n+2}}{1+x} .$$

The corresponding error bound for S_n is $|S_n - S_{n+1}| = \frac{x^{n+1}}{n+1}$, so this is a good theoretical improvement for each number x between 0 and 1. We report some calculations for the case $x = \frac{1}{2}$ in Table 11.3, ending our list with the correct value for $\ln(1+\frac{1}{2})$.

TABLE 11.3

Sum	Error
S_5 = 0.4072917	0.002
S_6 = 0.4046875	0.0008
T_5 = 0.4055556	0.00009
S_{10} = 0.4054346	0.00003
S_{11} = 0.4054790	0.00001
T_{10} = 0.4054642	0.0000009
$\ln 3/2$ = 0.4054651	

154

Clearly the practical results of this method are even greater than the theory predicts, and T_{10} is an acceptable method for calculating logarithms.

To recapitulate our method of refining the partial sums of an alternating series: we find a lower bound J_n for the absolute value of the remainder, $0 < J_n \leq |R_n|$, and define $T_n = S_n + (-1)^n J_n$. This implies that T_n lies between S_n and S and that the error in T_n is no greater than $|T_n - T_{n+1}|$.

EXAMPLE: REMAINDERS COMPARED TO GEOMETRIC SERIES

Consider the series $\sum_1 \frac{1}{i!}$ of positive terms: the remainder $R_n = \frac{1}{(n+1)!} + \frac{1}{(n+2)!} + \ldots$. To calculate each term a_{n+1} of this series, one may multiply the preceding term a_n by $\frac{1}{n+1}$. Hence after the first term $1/(n+1)!$ each term of R_n is no larger than the preceding one multiplied by $\frac{1}{n+2}$. Thus we may estimate R_n by the method of *comparison with a geometric series* with ratios $\frac{1}{n+2}$:

$$R_n \leq \frac{1}{(n+1)!} \sum_0 \left(\frac{1}{n+2}\right)^i = \frac{1}{(n+1)!} \frac{1}{1-\frac{1}{n+2}} = \frac{1}{(n+1)!} \frac{n+2}{n+1} .$$

Thus R_n is nearly as small as

$$a_{n+1} = \frac{1}{(n+1)!} ;$$

this means that our series converges nearly as rapidly as an alternating series.

Practically, this estimate for R_n implies that if we calculate (on an 8-digit machine) partial sums up to the point n where $(n+1)!$ is an 8-digit number, then S_n will be correct except possibly in its last digit.

155

You may instinctively respond, "Of course, if I add in all the terms that are large enough for my machine to call them non-zero, then I will surely get an answer that is accurate to the digital limits of my machine." But remember the harmonic series $\sum_1 \frac{1}{i}$: the nth term $1/n$ vanishes on an 8-digit calculator when n becomes larger than 10,000,000. The partial sum $S_{10,000,000} \leq S_{100} + 6S_{10} < 23$ (by a kind of reasoning discussed earlier in this chapter), yet we know that this series diverges so that there are partial sums as large as you please. Therefore, we do not, in general, have any guarantee that if the nth and subsequent terms vanish in the sight of our machine, S_n will be accurate to its digital capacity.

But for the series $\sum_1 \frac{1}{i!}$ we do have exactly that. Here are our calculations (remember, to find a_n, divide a_{n-1} by n):

11! = 39916800 and R_{10} = 0.0000003 so S_{10} will be accurate for us (see Table 11.4).

TABLE 11.4

n	S_n
5	1.7166667
9	1.7182815
10	1.7182818

It is easy to guess that our series has limit $S = e - 1$ and that S_{10} is indeed our best estimate.

ROUND-OFF

In addition to the truncation error R_n, there is another source of of error in the estimates of the value of any series $\sum_1 a_i$ by the calculation of a partial sum S_n. Each time a term a_i is calculated, its last digit may be in error, say by 1, due to *round-off*. This is the error caused by the machine's inability to display more than 8 (or 10 or whatever) digits, when a_i may be an infinite decimal. This round-off error may be increased by machine errors arising in the internal algorithms for irrational functions such as $\sqrt{x}$ and also by your use of an 8-digit, erroneous value for a_i in your computation of a_{i+1}. When n is 20, or even as small as 10, you could be so unlucky as to have all these errors in the same direction. They could then cause a total error affecting both of the last two digits of your answer, even though the remainder $|R_n|$ is much less than that. In our work, however, round-off will usually affect only the last digit.

EXERCISES

1. For each series indicated below, calculate its partial sum S_5 and state whether it converges or diverges.

*a. $\sum_1 1/2i$ d. $\sum_0 1/3^i$ *g. $\sum_1 (-1)^i/\sqrt{i}$

*b. $\sum_1 1/i^2$ e. $\sum_0 i/(2i-1)$ *h. $\sum_0 e^{-i}$

*c. $\sum_1 1/i(i+1)$ *f. $\sum_0 (i+1)/i!$ i. $\sum_1 \sin(1/i)/i$

*2. Euler showed that $\sum_1 1/i^2 = \sum_1 3(i-1)!^2/(2i)!$. That is, both series have the same sum $\pi^2/6$. Calculate the partial sums S_9 for each series.

3. For each of the sequences $2^i/i!$, $3^i/i!$, and $4^i/i!$, find the index i for the first term of the sequence that is less than 10^{-6}. Then give a proof that all of these sequences have limit 0.

*4. In Problem P2, Chapter 3 the limit of the sequence 67.89, $\sqrt{67.89}$, $\sqrt{\sqrt{67.89}}$, ..., $67.89^{1/2^n}$, ... was seen to be 1. More generally, $\lim_{n\to\infty} x^{1/n} = 1$ for every $x > 0$. To see that this is true, recall that $x^{1/n} = e^{\ln x/n}$ by definition. But $\lim_{n\to 0} \ln x/n = 0$, $e^0 = 1$, and e^y is a continuous function of y. Illustrate this fact by calculating and tabulating values for $x^{1/n}$ when $x = 0.25$. How large must n be in order to have $|0.25^{1/n} - 1| < 0.01$?

5. Follow the thread of Exercise 4, calculating a table to show that $\lim\limits_{n\to\infty} n^{1/n} = 1$. How large must n be in order that $|n^{1/n} - 1| <$ 0.01? Can you offer a proof that 1 is indeed the limit?

*6. Use the series $\ln(1+1) = 1 - 1/2 + 1/3 - \ldots = \sum\limits_{1} (-1)^{i+1}\, 1/i$

to calculate $\ln 2$. Display in your table the partial sums S_5, S_6, S_{10}, and S_{11} for this series, as well as the special sums T_5 and T_{10} that were discussed in our example of a remainder term for an alternating series. What is the error of each of your sums when compared to $\ln 2$?

7. John Wallis (1616-1703) showed that the *infinite product* $\dfrac{2X2X4X4X6X\ldots}{1X3X3X5X5X\ldots}$ had as its limit $\pi/2 = 1.5707963$. Calculate each partial product for this sequence out to $\dfrac{2X\ldots X14}{1X\ldots X13}$ and $\dfrac{2X\ldots X14}{1X\ldots X15}$, and then average these last two numbers for an estimate of $\pi/2$.

Next, take logarithms of the appropriate finite stages of this product to prove that it does indeed converge.

*8. The *Fibonacci sequence* 1, 1, 2, 3, 5, 8, 13, 21, 34, ... is formed by the rule $F_n = F_{n-1} + F_{n-2}$, with $F_1 = F_2 = 1$. The members of this sequence occur very frequently in phyllotaxis, or the study of the arrangement of leaves, scales of a pine cone, florets of a composite flower, and similar structures. That is, when k leaves are arranged in a staggered spiral that winds around a stalk n times, then k and n are quite likely to be Fibonacci numbers.

This sequence clearly diverges; we shall see that it does so in an orderly way. Show by calculation that the ratio $r_n = \dfrac{F_{n+1}}{F_n}$ of successive terms approaches a limit: $r_n \to 1.6180340$. Then use an arithmetic argument to show that $r_n = 1 + \dfrac{1}{r_{n-1}}$. Finally, give reasons why, if r_n approaches any limit, the limit must be the one given above. This number is called the *Golden Ratio*. A rectangle whose sides have this ratio is pleasing to the eye. The

Parthenon and many modern billboards are examples of man-made structures based on this ratio.

9. Estimate the remainder term R_{10} for the p-series $\sum_1 1/i^4 = \pi^4/90$
by comparison with an integral. Then use the upper and lower bounds
for R_{10} that you have found to form a new corrected sum T_{10}. Com-
pute T_{10} and compare the real error $|T_{10} - \pi^4/90|$ with the error
bound given in our worked-out example above for $\sum_1 1/i^2$. Also compare the real error of T_{10} with the real error of S_{20}.

10. Use the method of comparison with integrals to find upper and
lower bounds for the remainder term R_n for the series $\sum_{i=2}^{\infty} 1/i(\ln i)^2$
of positive terms. Then use these upper and lower bounds to define
a corrected sum T_n and to estimate the error for S_n and T_n. What
must n be in order that S_n, and then T_n, is accurate to five decimal
places (that is, a truncation error less than 5×10^{-6})?

11. Compare the remainder term R_{10} of the series $\sum_1 1/(i2^i) =$
0.6931472 (= ln 2) with a geometric series in order to show that it
satisfies

$$\frac{12}{11 \times 13 \times 2^{10}} < R_{10} < \frac{1}{11 \times 2^{10}}.$$

Use these bounds on R_{10} to form a corrected sum T_{10} and find the
real error for T_{10} as well as the computed error bound. Then compare the real error for T_{10} with that for S_{15}.

*12. Use a comparison with a geometric series to estimate R_{10} for the series $\sum_1 i/3^i$, finding both upper and lower bounds for the remainder. Then calculate a corrected sum T_{10} and estimate its error. It is easy to guess at the correct sum. Can you show why this is the sum of this series?

PROBLEMS

P1. Let the function $f(x)$ be defined by : $f(x) = \sum_1 (-1)^i \sin(ix)/i$ (this is called a *Fourier sin series for* f). From the definition we see that $f(x) = f(x+2\pi) = f(x+4\pi) = \ldots = f(x+2k\pi)$ for each integer k. Such a function is called *periodic with period* 2π; trigonometric functions are also periodic. In the present case, $f(\pi) = 0$ also.

Sketch graphs for the partial sums S_1, S_2, S_3, S_4. Then sum enough terms of the series to *convince yourself* of its values at the six points $\pi/4$, $\pi/2$, $3\pi/4$, $5\pi/4$, $3\pi/2$, and $7\pi/4$; use these values to sketch a graph for $f(x)$. (To read more about Fourier series, consult the text of Courant and John, which is cited in the Bibliography.)

P2. Illustrate the following theorems about the convergence of sequence by calculating the values when x has the indicated value and $n = 5$ or 10? Can you prove convergence in each case?

    a. $x^n \to 0$ if $|x| < 1$. Let $x = 0.99$.
    b. $x^n/n! \to 0$ for all x. Let $x = 4$.
    c. $1/n^x \to 0$ if $x > 0$. Let $x = 0.1$.
    d. $\ln n/n \to 0$.
    e. $n^x/e^n \to 0$ for all x. Let $x = 5$.

P3. Show that the sequence a_0, $a_0/\ln a_0$, $a_1/\ln a_1$, ..., $a_{n+1} = a_n/\ln a_n$, ... converges for every starting value $a_0 > 1$. As an aid to your thinking, experiment with the first five or six terms of this sequence for two different numbers a_0 of your own choosing. Then apply the graphic methods of Chapter 2 to the equations $y = x$ and $y = \dfrac{x}{\ln x}$.

P4. Calculate the values for n = 5, 10, 15, 20 in the sequence $n!e^n/n^{n+\frac{1}{2}}$ to illustrate that its limit is $\sqrt{2\pi}$. Then turn this expression around to regard it as a way of approximating $n!$ by a recipe involving e^n and n^n. This is called *Stirling's formula*. What is the error of your recipe for $n!$ when n = 5, 10, 15, or 20? (Hint: if your machine does not have scientific notation available, you must exercise some care in evaluating this sequence, lest the large intermediate numbers overflow the machine's capacity. There should be no problem if, for instance, the value for $n + 1$ is calculated by multiplying the value for n by $e\left(\dfrac{n}{n+1}\right)^{n+\frac{1}{2}}$. Consult the text of Courant and John or of James cited in the Bibliography for further information about Stirling's formula.)

*P5. Replace r in the geometric series by $-r^2$ to get

$$\frac{1}{1+r^2} = 1 - r^2 + r^4 - \ldots + (-1)^{n-1}r^{2n-2} + (-1)^n \frac{r^{2n}}{1+r^2} .$$

Then integrate both sides of this equality between 0 and x to get

$$\int_0^x \frac{dr}{1+r^2} = x - x^3/3 + x^5/5 - \ldots + (-1)^{n-1}x^{2n-1}/2n-1 + (-1)^n \int_0^x \frac{r^{2n}dr}{1+r^2} .$$

On the left-hand side we have the function arctan x. Show that if $|x| \leqq 1$, then the remainder term $R_n = \int_n^x \frac{r^{2n}dr}{1+r^2}$ has limit 0 as n tends to infinity. Thus the function arctan x may be approximated by the finite polynomials S_n or T_n. Illustrate this fact for x = 1 by calculating S_{10} and T_{10} for arctan 1 = $\pi/4$. Is this an efficient way to calculate $\pi/4$? What error bounds can be stated for S_{50}? For T_{50}?

*P6. We have seen that the harmonic series diverges. However, something may be said about the way in which it does so: the partial sums S_n grow about as fast as ln n grows! Specifically, Euler showed that there is a number $\gamma = 0.5772157$ (now called the *Euler number*) such that $\lim_{n \to \infty} (S_n - \ln n) = \gamma$. Calculate the values of the sequence $S_n - \ln n$ for $n = 10$ and $n = 50$. Though this convergence is quite slow, you may wish to pursue it to $n = 100$ on your machine. In each case, compute the error.

*P7. In the geometric series, replace r by $-r^3$ and integrate both sides of the resulting equality from 0 to 1 to obtain

$$\int_0^1 (1/1+r^3)dr = r - r^4/4 + r^7/7 - \ldots + (-1)^{n-1}r^{3n-2}/3n-2 + \ldots \Big]_0^1 .$$

Show that the remainder after summing n terms is

$$R_n = \int_0^1 (-1)^n \left[r^{3n}/(1+r^3) \right] dr$$

and that

$$\lim_{n \to \infty} R_n = 0.$$

Then integrate the function $1/(1+r^3)$ from 0 to 1 (by partial fractions or otherwise) to obtain

$$0.8356488 = 1 - 1/4 + 1/7 - 1/10 + \ldots .$$

Finally, establish bounds on R_n to define a correction term for S_n and calculate S_{10} and T_{10}. Display your results together with their estimated and actual errors.

P8. Develop a power series for $\ln(1-x)$ when $0 \le x < 1$. Do this by following our Example, which expressed $\ln(1+x)$ as an alternating power series, down to the stage where the assumption was *used* that $x > 0$. Make new remainder estimates for your case, prove that your

162

series converges for each number x between 0 and 1, and define a corrected partial sum. (Caution: your series will have all its terms negative — treat this series just as you would a series of positive terms.)

Use your corrected sum T_{10} to estimate $\ln(1-1/3)$ and $\ln(1-9/10)$. Finally, show that your series, for $0 \leqq x < 1$ and $\ln(1-x)$, may be placed alongside the series of the Example, for $0 \leqq x \leqq 1$ and $\ln(1+x)$, to show that the series of the Example in fact converges for $-1 < x \leqq 1$.

P9. Suppose we seek a *continued fraction* expression for a real number x, so

$$x = a_0 + \cfrac{1}{a_1 + \cfrac{1}{a_2 + \ddots}}$$

or $x = a_0 + \cfrac{1}{a_1+}\ \cfrac{1}{a_2+}\ \ldots$. The numbers a_i in this expression are to be integers; such a continued fraction is often called *simple*. (See also Problems P5, Ch. 6 and P3, Ch. 8.) Let us denote by $[x]$ the *greatest integer in* x; that is, $[x]$ is x minus its fractional or decimal part. Then we may take a_0 to be $[x]$; if we define $y_i = a_i + \cfrac{1}{a_{i+1}+}\ \cfrac{1}{a_{i+2}+}\ \ldots$, then $x = y_0$ and also $x = a_0 + 1/y_1$. Hence $y_1 = 1/(y_0-a_0)$; we choose $a_1 = [y_1]$. In general, $y_i = a_i + 1/y_{i+1}$, so we may choose $a_i = [y_i]$ and compute $y_{i+1} = 1/(y_i-a_i)$. This is a very rapid process on a calculator. For instance, we calculate the continued fraction for $x = 17/11 = 1.5454545$:

$$x = y_0 = 1.545454$$
$$a_0 = 1.$$
$$y_0 - a_0 = 0.545454$$
$$y_1 = 1.8333333$$
$$a_1 = 1.$$
$$y_1 - a_1 = 0.8333333$$
$$y_2 = 1.2$$
$$a_2 = 1.$$
$$y_2 - a_2 = 0.2$$
$$y_3 = 5.$$
$$a_3 = 5.$$

Hence $17/11 = 1 + \dfrac{1}{1+} \dfrac{1}{1+} \dfrac{1}{5}$ (check this!). It is easy to see that a finite, terminating continued fraction is a rational number. The converse is true as well: every rational number has a finite continued fraction expansion. Do you see why this process must terminate if x is rational?

Calculate the continued fractions for $x = \sqrt{2}$ and $x = \sqrt{63}$.

Show that $3 + \dfrac{1}{7+} \dfrac{1}{16}$ is a good approximation to $\pi = 3.1415927$ by using the above process. The successive, rationals a_0, $a_0 + \dfrac{1}{a_1}$, $a_0 + \dfrac{1}{a_1+} \dfrac{1}{a_2}$, ... are called the *convergents* of the continued fraction, so the above problem may be restated as: Show that 3, 22/7, 333/106, 355/113 are the first four convergents for π.

Next, find the continued fraction expansion and the convergents for $e = 2.7182818$, for the Golden Ratio $(1+\sqrt{5})/2$ (see Exercise 8), and also for the *Euler number* $\gamma = 0.5772157$ (γ is described in Problem P6). Calculate the error for your convergent at each stage and halt the process when round-off error has accumulated to cause the calculated convergents to cease converging toward e or γ. Label your best convergent in each case; is it a good rational approximation?

To read more about continued fractions, look in a book about "number theory."

P10. The *Fibonacci sequence* is the sequence of integers 1, 1, 2, 3, 5, ..., where $F_1 = F_2 = 1$ and $F_n = F_{n-1} + F_{n-2}$ (see Exercise 8). Show that there are numbers r and s such that

$$F_{n+1} - r F_n = s(F_n - r F_{n-1}),$$
$$F_{n+1} - s F_n = r(F_n - s F_{n-1}).$$

Hence that we may write

$$F_{n+1} - s F_n = r^n,$$
$$F_{n+1} - r F_n = s^n.$$

Next, subtract to get the formula

$$F_n = 1/\sqrt{5}\left[\left(\frac{1+\sqrt{5}}{2}\right)^n - \left(\frac{1-\sqrt{5}}{2}\right)^n\right],$$

which expresses F_n in "closed form," so that it may be calculated directly, without the intermediate computation of F_1, F_2, F_3, ..., F_{n-1}. Use this formula to prove that the limit of Exercise 8, $\lim_{n\to\infty} F_{n+1}/F_n$, does indeed exist.

Finally, prove that if we adopt the notation that $< x >$ is the nearest integer to x (rounding upward from halves), then

$$F_n = \left\langle \frac{1}{\sqrt{5}}\left(\frac{1+\sqrt{5}}{2}\right)^n \right\rangle.$$

The function $< x >$ may be obtained on calculators that can fix the number of decimal places. If your machine will fix its display at zero decimal places *with* rounding-up, then that is F_n. If your machine will display its result with no digits after the decimal,

165

without rounding-up, then first add ½, to display

$$F_n = \left[\frac{1}{\sqrt{5}} \left(\frac{1+\sqrt{5}}{2} \right)^n + \frac{1}{2} \right]$$

without its decimal part. To read more about Fibonacci sequences, consult texts about "number theory."

P11. If a_0, a_1, ... is a sequence, the function $f(x) = \sum_0 a_i x^i$ is called the *generating function* for the sequence. Show that $1/(1-x-x^2)$ is the generating function for the Fibonacci sequence (see Exercise 8 and Problem P10). Then calculate S_{10} for $\sum_0 F_i 10^{-i}$ and compare this partial sum to the infinite sum computed with the generating function.

Answers to Starred Exercises and Problems

Exercises   1a. **1.1416667**, diverges

                 1b. **1.4636111**, converges

                 1c. **0.8333333**, converges

                 1f. **5.435**, converges

                 1g. **-0.8174571**, converges

                 1h. **1.5780554**, converges

                  2. **1.5397677** and **1.6449339**, respectively

                  4. $n = 2^8$

                  6. $S_5 = $ **0.7833333**, error is **0.09**

                       $S_{10} = $ **0.6456349**, error is **0.05**

                       $T_5 = $ **0.7**, error is **0.007**

                       $T_{10} = $ **0.6910895**, error is **0.002**

                  8. $r_n \to (1+\sqrt{5})/2 = \varphi$; $\varphi - 1/\varphi = 1$

                 12. $11/2 \text{X} 3^{10} < R_{10} < 121/7 \text{X} 3^{11}$

                       $S_{10} = $ **0.7499026**, $T_{10} = $ **0.7499980**

                       and $S = 3/4$

Problems   P5. $S_{50} = 0.7803987$ with 0.6% error

           P6.   The error is 0.05 for $n = 10$, 0.01 for $n = 50$, 0.005 for $n = 100$.

           P7.   The integral is $(\pi/\sqrt{3} + \ln 2)/3$.

12

POWER SERIES

INTRODUCTION

This chapter continues our study of series. We shall now extend the usefulness of series methods enormously by exploiting the notion basic to the power series, which we have already seen. This is the idea of a series of functions, a series each term of which is a multiple of a power of x. This chapter begins with three theorems that methodically describe the convergence and manipulation of such series. We apply these theorems to the exponential function and continue our study by attempting to approximate e^x with polynomials. This leads to Taylor's theorem and its remainder term, which are again realized for the Example e^x.

The first few Exercises develop numerical and calculational skills with power series. Then series for sin x, cos x, sinh x, cosh x, and 10^x are presented in further Exercises; and their values are calculated. In the final Exercise we will consider the biological question of averaging exponential growth rates. In the Problems section Taylor expansions are obtained for $(x+1)e^x$, $e^{\cos x}$, $\sqrt{1+\sin x}$,

$(1+x)^{\alpha}$, arcsin x, $x/(e^x-1)$, tan x, and a rational function. In one Problem we present a tricky trig identity for a new, easy, and accurate calculation of π. Series calculations of the Bernoulli and Euler numbers are also Problems as is a Padé approximation.

THE THEOREMS

We saw in the last chapter some examples of power series: the geometric series $\frac{1}{1-x} = 1 + x + x^2 + \ldots$ and $\ln(1+x) = x - \frac{x^2}{2} + \frac{x^3}{3} - \ldots$ in our worked-up Examples (and also arctan $x = x - \frac{x^3}{3} + \frac{x^5}{5} - \ldots$ in Problem P5, Ch. 11). In the first case the function $\frac{1}{1-x}$ is defined for every $x \neq 1$, yet $\sum_0 x^i$ converges only for $-1 < x < 1$, where the series does converge to the value $\frac{1}{1-x}$ of the function. The function $\ln(1+x)$ is defined for all $x > -1$, yet the series converges only for $-1 < x \leq 1$. These two examples will illustrate some general remarks, which we record as THEOREMS.

1. *A power series* $\sum_0 a_i x^i$ *may converge for every* x, *or it may converge only for* x = 0. *Otherwise there is a definite positive number* r, *the radius of convergence, such that the series diverges when* $|x| > r$ *and converges whenever* $|x| < r$. *If the limit* $\lim_{n\to\infty} \sqrt[n]{a_n}$ *exists, then it equals* 1/r. *The series may or may not converge at either endpoint* -r *or* r *of its interval of* convergence.

2. *If we write* S(x) = $\sum_0 a_i x^i$, *with radius of convergence* r, *then for every* x *inside the interval of convergence,* $|x| < r$, *the series* $\sum_0 i a_i x^{i-1}$ *of derivatives of the terms of* S(x) *converges to the derivative* S'(x). *Also the series* $\sum_0 a_i x^{i+1}/(i+1)$ *of integrals of the terms converges to* $\int_0^x$ S(x)dx. *Hence, inside its interval of convergence* S(x) *is continuous and in fact has derivatives of all orders. If the series converges at an endpoint of the interval of convergence, then* S(x) *is continuous at that point (in the one-sided sense).*

3. *If there exists an* r > 0 *such that two series converge and*
$$\sum_0 a_i x^i = \sum_0 b_i x^i \text{ for every x with } |x| < r, \text{ then } a_i = b_i \text{ for every}$$
index i = 1, 2, 3, ..., *so the two series are identical.*

EXAMPLE: e^x

As an example of the use of these facts, we consider the series
$E(x) = \sum_0 x^i/i!$. Since $\lim_{n \to \infty} x^n/n! = 0$ for every x, the alternating
series $E(-x)$ converges for every positive x. Hence $E(x)$ converges
for all x, by the first theorem above. (Be sure you understand this!)
Incidentally, that theorem may now be reread in this case to give
the interesting information that $\lim_{n \to \infty} \sqrt[n]{1/n!} = 0$. The derivative $E'(x)$
is, by the second theorem, $E'(x) = \sum_0 i x^{i-1}/i! = \sum_0 x^i/i! = E(x)$ for
all x. Thus the function $f(x) = \ln E(x)$ has derivative $\frac{E(x)}{E(x)} = 1$ for
all x, and $f(0) = 0$. And we know that there is a unique solution to
the problem of finding an antiderivative f for f', given that $f'(x) = 1$ and $f(0) = 0$. Namely, $f(x) = x$, or $x = \ln E(x)$, and thus $e^x = E(x)$.

Therefore the series $E(x) = 1 + x + \frac{x^2}{2!} + \frac{x^3}{3!} + \ldots = e^x$ is the
unique series representing the exponential function, by the third
theorem above, and this representation is valid for every real number
x. We pause to use this series to calculate

$$e^{-1} = 1/e = 1 - 1 + \frac{1}{2!} - \frac{1}{3!} + \ldots,$$

which is an alternating series with remainder $|R_n|$ less than $\frac{1}{(n+1)!}$,
the size of the first omitted term after S_n. When $n = 9$, we have
$|R_9| < 1/10! = .0000003$ and $S_9 = 0.3678792$; the correct sum is $1/e = 0.3678794$. (If your machine does not have a button for $i!$, notice
that each term is easy to calculate from the preceding one, $a_i = \frac{-1}{i} a_{i-1}$, after beginning with $a_2 = 1/2$. See the Appendix for other
tricks useful in evaluations.)

170

Taylor Polynomials

The statement that $e^x = \sum_0 x^i/i!$ may be viewed as a fact about the approximation of the function e^x by various polynomial functions S_n. The function e^x is approximately equal to $x^2/2 + x + 1$, for instance, with a better approximation given by $e^x \doteq x^3/6 + x^2/2 + x + 1$, or even by $e^x \doteq S_{10}(x) = x^{10}/10! + x^9/9! + \ldots + x + 1$. And clearly, polynomials are desirable functions with which to approximate e^x, since we can easily calculate the value of a polynomial using only addition, subtraction, multiplication, and division. But, in what sense are these good approximations?

In approximating a number, the error is a number, and the better the approximation, the smaller the error. But in approximating e^x by any polynomial $P(x)$ whatsoever, it is easy to see that as x gets larger, the error $|e^x - P(x)|$ becomes larger without limit. (Can you say why?) In what way, then, may we regard $x^2/2 + x + 1$ as *the best quadratic approximation* to e^x? We have something that we know already to go on: the *best linear approximation* at a given point, say $x = 0$, is given by the derivative. That is, the best linear approximation to e^x at $x = 0$ is the straight line with slope m equal to the derivative of e^x at 0 and that goes through the point $(0, e^0)$. Since $m = e^0 = 1$, this tangent line has the equation $y = x + 1$. It is the best we can do to fit the graph of e^x at $x = 0$ with a line (Figure 12.1). We could describe $x + 1$ as the *unique* polynomial of degree one that has the same value at $x = 0$ as the function e^x and also has the same first derivative.

Figure 12.1

Now we go back to inspect $x^2/2 + x + 1$: it is the unique quadratic polynomial that agrees with e^x at $x = 0$ in its value, its first derivative, and also its second derivative. This means that the graph of $x^2/2 + x + 1$ not only touches the graph of e^x at $x = 0$ and is tangent to e^x there; it also has the same curvature there as e^x. This is shown in Figure 12.2. And this is the sense in which we regard it as the best quadratic approximation to e^x.

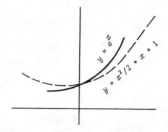

Figure 12.2

The choice of $x = 0$ is arbitrary; we might have discussed the approximation of e^x at $x = 4$, for instance, instead of $x = 0$, and we will do so in the next chapter. The essential concept is the agreement of two functions at a point, in value of functions, of first derivatives, of second derivatives, and so on. The English mathematician Brook Taylor saw this possibility in 1712 and used it to develop a powerful method of approximating functions and of expressing functions as power series.

Suppose the polynomial $S_n(x) = a_0 + a_1x + \ldots + a_n x^n$ agrees with a function $f(x)$ at the point $x = 0$; then $a_0 = f(0)$. (Notice here that there is a term corresponding to $i = 0$, so that $S_n(x) = \sum_0 a_i x^i$.) If $S_n'(x) = a_1 + 2a_2x + 3a_3x^2 + \ldots + na_n x^{n-1}$ agrees with $f'(x)$ at $x = 0$, then $a_1 = f'(0)$. Another such step shows that $a_2 = f''(0)/2$ and that in general $S_n^{(i)}(0) = i!a_i = f^{(i)}(0)$ for $i = 0$, 1, $\ldots$, n. Hence the proper coefficients for the *Taylor polynomial* are $a_i = f^{(i)}(0)/i!$ and

$$S_n(x) = f(0) + f'(0)x + f''(0)x^2/2! + f'''(0)x^3/3! + \ldots + f^{(n)}(0)x^n/n!.$$

THE REMAINDER FUNCTION

Furthermore, we know a lot about the remainder or error function

$$R_n(x) = f(x) - S_n(x): \quad R_n(0) = R_n{}'(0) = R_n{}''(0) = \ldots = R_n{}^{(n)}(0) = 0,$$

and all derivatives of $R_n(x)$ of order greater than n agree with the derivative of same order of f. This description of $R_n(x)$ can be used to obtain an integral representation of the remainder when $f^{(n+1)}(t)$ is continuous on the interval $[0,x]$:

$$R_n(x) = \frac{1}{n!} \int_0^x (x-t)^n f^{(n+1)}(t) \, dt.$$

The statement that f(x) *differs from its Taylor polynomial by this remainder term,* f(x) = S_n(x) + R_n(x), *for each integer n for which* $f^{(n+1)}$(x) *is continuous* is called TAYLOR'S THEOREM. However, it is usually not useful in this form because this integral is seldom easy to evaluate.

We can estimate the integral, though, using upper and lower bounds $L \le f^{(n+1)}(x) \le M$ for the continuous function $f^{(n+1)}$ on the interval $[0,x]$:

$$\frac{L}{n!} \int_0^x (x-t)^n \, dt \le R_n(x) \le \frac{M}{n!} \int_0^x (x-t)^n \, dt$$

$$\frac{L}{n!} \frac{x^{n+1}}{n+1} \le R_n(x) \le \frac{M}{n!} \frac{x^{n+1}}{n+1}$$

$$\frac{L}{(n+1)!} x^{n+1} \le R_n(x) \le \frac{M}{(n+1)!} x^{n+1}.$$

Since the continuity of the function $f^{(n+1)}(x)$ implies that it takes on every value between L and M, there is a number ξ between 0 and x such that

$$R_n(x) = \frac{f^{(n+1)}(\xi)}{(n+1)!} x^{n+1}.$$

173

This is called *Lagrange's form of the remainder*. It is this expression and the equivalent inequality that precedes it that will give us useful estimates of $R_n(x)$. For instance, for the function $f(x) = e^x$, which we were examining, $S_n(x) = 1 + x + x^2/2 + \ldots x^n/n!$; the remainder $R_n(x) = \frac{e^\xi}{(n+1)!} x^{n+1}$. Of course, if we knew the values of e^ξ, we wouldn't need to approximate this function; however we can estimate that $e^\xi < 3^X$, where X is the smallest integer at least as large as x. We emphasize that accuracy is not essential in this estimation. We can be assured of the maximal size of the error if we are in the ballpark in our estimate of $f^{(n+1)}(\xi)$.

EXAMPLE: THE CALCULATION OF e^x

Thus if we wish to calculate $e^{1.7}$ correct to 5 decimal places, we take $X = 2$ so $R_n(1.7) < \frac{3^2}{(n+1)!} (1.7)^{n+1}$. Next we find the first integer n for which $R_n(1.7) < 0.000005$: we see that $n = 12$ since $R_{12}(1.7) < \frac{3^2}{(13)!} (1.7)^{13} = 0.0000014$. Thus we are to calculate $S_{12}(1.7) = 5.4739472$; the last term we added was $x^{12}/12! = .0000012$, which is just about the size of $R_{12}(1.7)$. The correct value $e^{1.7} = 5.4739474$, so that our result was in fact accurate in the sixth decimal place. One more term would make $S_{13}(1.7)$ correct in all 8 digits. Incidentally, if your machine does not have a buttton to calculate $n!$, you can calculate each term from the preceding one as before, $a_i x^i/i! = \frac{x}{i} (a_{i-1} x^{i-1})/(i-1)!$, summing as you go.

EXAMPLE: ALTERNATIVE METHODS FOR e^x

The error term $R_n(x) = e^\xi x^{n+1}/(n+1)!$ depends on x, of course, and it is clearly a lot smaller for $x/2$ than for x. But $(e^{1.7/2})^2 = e^{1.7}$, so we could calculate $e^{0.85}$ using fewer terms than for $e^{1.7}$, then square the result. $R_n(0.85) < 3(0.85)^{n+1}/(n+1)!$, so $n = 8$ is large enough to guarantee five decimal places of accuracy.

Encouraged by this, let us repeat the process twice more, dividing 1.7 by 8 to examine the calculation of $e^{0.2125}$. The error

174

term is now $R_n(0.2125) < 3^{\frac{1}{4}}(0.2125)^{n+1}/(n+1)!$, and this latter quantity is less than 0.000005 when $n = 4$. So to calculate $e^{1.7}$, compute $S_4(1.7/8)$ and take the eighth power of the result. This gives $e^{1.7} \doteq 5.4738149$, which is incorrect in its fourth decimal place. What has gone wrong? Well, we didn't allow for the error resulting from the final operation, raising to the eighth power. If $g(x) = x^8$, then $dg = 8x^7 dx$; since $x = e^{0.2125} \doteq 5/4$, $dg \doteq 35dx$. Thus we shall need at least 6-place accuracy for $e^{0.2125}$ to insure 5 correct decimal places in its eighth power $e^{1.7}$. It will suffice to add one more term, to calculate $S_5(0.2125) = 1.2367660$. Its eighth power is $e^{1.7} \doteq 5.4739427$, which is correct through its fifth decimal place.

We summarize the computational experience above: our series expansion for e^x about the point $x = 0$ provides a given degree of accuracy with fewer terms of the series when x is closer to 0. Thus we need fewer terms to calculate $S(x/2)$ than to calculate $S(x)$, correct to five places, say. But there is an increase in error when we square $S(x/2)$ to get $S(x)$. On balance, ease of computation favors several halvings of x before calculating $S(x/4)$, $S(x/8)$, etc. But for accuracy nearing the limit of the machine, we must do more additions, computing $S(x)$ directly or from $S(x/2)$.

EXERCISES

1. For each series indicated here, evaluate the partial sum of the first four terms when $x = 0.56789$. (Consult the Appendix, if necessary, to develop techniques for summing power series efficiently on your machine.)

*a. $1 - x^2/2! + x^4/4! - \ldots$ *c. $x - x^3/3! + x^5/5! - \ldots$

*b. $x - x^3/3 + x^5/5 - \ldots$ *d. $x - x^4/4 + x^7/7 - \ldots$

*2. Use the series $e^x = 1 + x + x^2/2! + x^3/3! + \ldots$ to calculate $e^{-0.1}$ correct to five decimal places. Do this by summing terms for your *alternating* series until the last term to be summed is smaller than 5×10^{-6}.

3. Proceed as in Exercise 2 to calculate $e^{-0.2}$.

175

*4. Proceed as in Exercise 2 to calculate $e^{-0.4}$. Compare your answer to the correct value and to the 4th power of your answer to Exercise 2.

5. Proceed as in Exercise 2 to calculate $e^{-0.8}$. Compare your answer to the 8th power of your answer to Exercise 2.

*6. The sequence of derivatives $f(x)$, $f'(x)$, $f''(x)$, ... for $f(x) =$ sin x is sin x, cos x, -sin x, -cos x, sin x, ... The coefficients of the Taylor polynomials for sin x are these functions evaluated at $x = 0$: 0, 1, 0, -1, 0, ... Thus

$$\sin x = x - x^3/3! + x^5/5! - x^7/7! + \ldots,$$

with remainder term $R_n(x) = \dfrac{\sin^{(n+1)}(\xi)}{(n+1)!} x^{n+1}$. Since $|\sin x| \leq 1$ and also $|\cos x| \leq 1$ for all x, we have $|R_n(x)| \leq x^{n+1}/(n+1)!$. Find the appropriate integer n for which $S_n(0.1)$ will have five-place accuracy and calculate S_n. Notice that n is odd, so $S_{n+1} = S_n$ and the real error is less than $|R_{n+1}(0.1)|$. Calculate that better error bound and compare it to the error for your answer. (Remember that radian measure is meant for x.)

7. Proceed as in Exercise 6 to calculate sin 1 accurate to five places. Then calculate sin $1°$ (degree measure !) to five-place accuracy.

*8. Proceed as in Exercise 6 to calculate sin 3 correct to five decimal places. Then use the trig identity sin $x = $ sin $(\pi-x)$ and the same method to get a five-place answer. Compare the number of terms in the two partial sums.

9. Proceed as in Exercise 6 to show that

$$\cos x = 1 - x^2/2! + x^4/4! - x^6/6! + \ldots .$$

Now compute sin 3/2 by using the identity sin $x = $ cos $(x-\pi/2)$, find the least integer n and the value $S_n(x-\pi/2)$ for five-place accuracy.

What is the appropriate value for n to compute sin 3/2 to five places? Can you offer a new proof now that the derivative of sin x is cos x?

*10. The *hyperbolic sine* is the function sinh $x = \frac{1}{2}(e^x - e^{-x})$. The *hyperbolic cosine* is cosh $x = \frac{1}{2}(e^x + e^{-x})$. Show that each of these functions is the derivative of the other, and then use this fact to establish the following coefficients for their Taylor polynomials:

$$\sinh x = x + x^3/3! + x^5/5! + \ldots ,$$
$$\cosh x = 1 + x^2/2! + x^4/4! + \ldots .$$

Evaluate the appropriate partial sum for each of these expressions to find sinh 1/2 and cosh 1/4 correct to 5 decimal places.

11. The *antilog* function 10^x may be evaluated by appeal to its definition, $10^x = e^{x \ln 10}$, followed by evaluation of the appropriate Taylor polynomial for the exponential function. Prove that the expansion of the function 10^x into its own Taylor polynomial offers no improvement on this; in fact it yields the same method. Then use the series for e^x to calculate $10^{0.3}$. In doing so you will need to know ln 10. To compute this most easily, first calculate ln 5/4 using the expression

$$\ln(1+x) = x - x^2/2 + x^3/3 - \cdots$$

and then use our previously determined value ln 2 = 0.6931472 to obtain $\ln(2^3 \times 5/4)$. (Why not directly calculate $\ln(1+9)$?) Be sure when you are deciding how many terms of the series for e^x to use that you consider the ultimate error in 10^x, not just the error of e^x.

12. Suppose that the larger part of the claw of a crab grows exponentially at the rate of 9% per month in weight, while the smaller part grows at the rate of 7% per month (compare Exercise 13, Ch. 8). Assume that at time $t = 0$ months the smaller part is one-half the weight of the larger. Does the whole claw grow exponentially? First answer this question by calculating the rate r that would

177

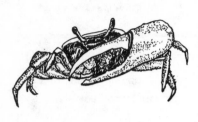

satisfy $e^r = 2/3 \ e^{0.09} + \frac{1}{3} e^{0.07}$ (at
the end of $t = 1$ month). Then check
whether the appropriate equality holds
at the end of 2 months.

Finally, use the quadratic
Taylor polynomials for the respective
functions to show that such a (nontrivial) proportional sum of exponential functions is never an exponential
function.

Problems

*P1. Find the sequence of Taylor polynomials for the function $f(x) =$
$(x+1)e^x$. Choose the one of these of least degree to estimate $f(1)$
correct to five decimal places and calculate that estimate.   Then
show that your sequence converges for every number x to $f(x)$.

P2. Compute the coefficients for the Taylor polynomial $S_6(x)$ of
degree 6 for $f(x) = e^{\cos x}$; then calculate $S_6(0.1)$ and $S_6(\pi/4)$.   For
each of these approximations, make an estimate of the maximal error.

*P3. Compute the coefficients for the Taylor polynomial $S_6(x)$ of
degree 6 for $f(x) = \sqrt{1 + \sin x}$; then calculate $S_6(0.1)$ and $S_6(\pi/4)$.
For each of these approximations, make an estimate of the maximal
error.

P4. Prove that, for each real number α:

$$(1+x)^\alpha = 1 + \alpha x + \alpha(\alpha-1)x^2/2! + \alpha(\alpha-1)(\alpha-2)x^3/3! + \dots$$

whenever $|x| < 1$. Do this by first establishing the interval of con-
vergence of this series by use of the ratio test (or comparison with
a geometric series). Next, show that in the interval of convergence
where $S(x)$ is the sum of the series, $S'(x) = \alpha S(x)/(1+x)$ and also
$S(0) = 1$. This means that the function $\ln S(x) = T(x)$ has the properties $T'(x) = \alpha/(1+x)$ and $T(0) = 0$. Argue finally that there could
be at most one function having these properties of $T(x)$ and that

178

$T(x) = \alpha\ln(1+x)$ is such a function.

Now use your series to calculate $\sqrt[3]{1.2}$. Analyze the error to decide how many terms are necessary for a partial sum to be accurate to five decimal places and compute that partial sum.

P5. Use trig identities to prove that if $\tan \theta = 1/5$, then $\tan(4\theta-\pi/4) = 1/239$. The formulas for tangents of sums and differences of angles will express $\tan 2\theta$, $\tan 4\theta$, and then $\tan(4\theta-\pi/4)$. Finally, show that

$$\pi = 16 \arctan 1/5 - 4 \arctan 1/239.$$

Calculate π correct in at least six decimal places by means of this recipe and the series we have derived for arctan x, you will need to make your total error term less than 5×10^{-7}. In doing this job, first establish that the remainder after only *one* term for arctan $1/239$ will be acceptable. Next multiply this remainder by 4 and subtract the result from 5×10^{-7} to establish the allowable error in $16 \arctan 1/5$. Then calculate the appropriate partial sum for arctan $1/5$. (If your machine displays ten digits, you may do this approximation to eight-digit accuracy. The above method has been used on a computer to achieve 100,000 digit accuracy!)

P6. Use the result of Problem P4, the series for $(1+x)^{\alpha}$, to prove that

$$\arcsin x = \int_0^x \frac{dt}{\sqrt{1-t^2}} = x + \frac{1 \times x^3}{2 \times 3} + \frac{1 \times 3 \times x^5}{2 \times 4 \times 5} + \frac{1 \times 3 \times 5 \times x^7}{2 \times 4 \times 6 \times 7} + \dots .$$

Then calculate the appropriate partial sum for arcsin (0.3), accurate to five places.

P7. Leonhard Euler proved in 1731 that

$$\sum_1 1/i^2 = (\ln 2)^2 + 2 \sum_1 1/i^2 2^i.$$

179

We saw in Chapter 11 that $\sum_1 1/i^2$ converges very slowly indeed to $\pi^2/6$; fifty terms do not suffice for two-place accuracy. We did achieve four-place accuracy in our earlier discussion with the corrected sum T_{20}.

Calculate the partial sum S_{20} for $\sum_1 1/i^2 2^i$. Next make an error estimate and attempt to improve your sum S_{20} with a correction term that you develop from your study of the remainder.

P8. Write $\dfrac{x^2}{a^4-x^4}$ as a power series; do so by first expressing this rational function in terms of partial fractions and then finding a series for each fraction (for the series of $\dfrac{1}{1+r^2}$ see Problem P5, Ch. 11).

P9. Expand the function $x \cot x$ into its Taylor series, deriving the first eight terms. These terms are often expressed as $x \cot x =$ $1 - \dfrac{2^2 B_2}{2!} x^2 + \dfrac{2^4 B_4}{4!} x^4 + \ldots + (-1)^n \dfrac{2^{2n} B_{2n}}{(2n)!} x^{2n} + \ldots$, where the numbers B_2, B_4, ... are called the *Bernoulli numbers*. Thus, show that $B_2 = 1/6$, $B_4 = -1/30$, $B_6 = 1/42$, $B_8 = -1/30$, ...

Also, $B_1 = -1/2$ and $B_3 = B_5 = B_7 = \ldots = 0$, but this series does not make at least the first of these odd Bernoulli numbers explicit. Show that the series for $x/(e^x-1)$ begins as

$$1 - x/2 + B_2 x^2/2! + B_3 x^3/3! + \ldots .$$

Finally, use the trig identity $\tan x = \cot x - 2 \cot 2x$ to show that $\tan x = x + x^3/3 + 2x^5/15 + 17x^7/315 + \ldots$ and express each summand of this series in terms of the Bernoulli numbers.

*P10. The Bernoulli numbers (which are defined in Problem P9) may be computed from series. For $n = 1, 2, \ldots,$

$$B_{2n} = \dfrac{(-1)^{n-1} 2(2n)!}{(2\pi)^{2n}} \sum_1 \dfrac{1}{i^{2n}} .$$

Use this expression to calculate B_{10} and B_{12}.

*P11. The *Euler numbers* E_1, E_2, ... are defined as the coefficients in the series

$$\frac{2^{n+1}e^{x/2}}{e^x+1} = \sum_1 E_n \, x^n/n! \ .$$

Show from this definition that the odd Euler numbers E_1, E_3, ... are all zero, and that $E_0 = 1$, $E_2 = -1$, $E_4 = 5$.

Next, use the series expression for the Euler numbers,

$$E_{2n} = \frac{(-1)^n 2^{2n+2}(2n)!}{\pi^{2n+1}} \sum_0 \frac{1}{(2i+1)^{2n+1}} \ ,$$

to calculate E_6 and E_{12}. (It is true that E_n is always an integer.)

P12. Calculate the limit

$$\lim_{h\to 0} \frac{\tan h - \sin h}{h^3} \ .$$

Then use the series expressions for sin and tan (see Problem P9 for tan) to prove that your calculated limit is correct.

P13. Suppose we attempt to find a rational function

$$R(x) = \frac{A + Bx + Cx^2}{D + Ex}$$

that approximates e^x near $x = 0$. To have $R(x)$ defined at all at $x = 0$ we must choose $D \neq 0$; we "normalize" the problem by taking $D = 1$.

Now we wish $\dfrac{A + Bx + Cx^2}{1 + Ex}$ to equal $e^x = 1 + x + \dfrac{x^2}{2} + \dfrac{x^3}{6} + \ \cdots$

or

$$A + Bx + Cx^2 = (1+Ex)\left(1 + x + \frac{x^2}{2} + \frac{x^3}{6} + \ldots\right)$$

$$A + Bx + Cx^2 + 0x^3 = 1 + (1+E)x + \left(\frac{1}{2} + E\right)x^2 + \left(\frac{1}{6} + \frac{E}{2}\right)x^3 + \ldots \; .$$

If we take $\frac{1}{6} + \frac{E}{2} = 0$ or $E = -1/3$, then the coefficient of x^3 on the right-hand side will be 0. This, in turn, implies that $A = 1$, $B = 2/3$, and $C = 1/6$, and

$$R(x) = \frac{1 + 2x/3 + x^2/6}{1-x/3} = \frac{6 + 4x + x^2}{6 - 2x} \; .$$

Calculate the error in $R(x)$ for $x = 0.01$, 0.1, and 1.

A Padé approximation of degree (m,n) to a function $f(x)$ at $x = 0$ is a rational function, a quotient $P_m(x)/Q_n(x)$ of two polynomials of degrees m and n respectively. This approximation $P_m(x)/Q_n(x)$ is to agree with $f(x)$ and its first $m + n$ derivatives at $x = 0$. Let $Q_n(x) = q_0 x^n + q_1 x^{n-1} + \ldots + q_n$; since $Q_n(0)$ cannot be 0, we normalize by taking $q_n = 1$. If $P_m(x) = p_0 x^m + \ldots + p_m$, then we have available $m + n + 1$ independent choices of the coefficients p_0, $\ldots$, p_m, q_0, $\ldots$, q_{n-1}.

These choices may be made so that if $f(x) = \sum_0 a_i x^i$, then $P_m(x)$ and $Q_n(x) \sum_0 a_i x^i$ have the same coefficients for all terms of degree $\leq m + n$. Find the Padé approximations of degrees $(3,2)$ and $(4,3)$ for $f(x) = \sin x$, and determine the errors for each approximation when $x = 0.01$, 0.1, and 1.

Answers to Starred Exercises and Problems

Exercises   1a.   0.8430375   1c.   0.5378544

            1b.   0.5159336   1d.   0.5442609

            2. $S_4 = 0.9048375$; $e^{-0.1} = 0.9048374$

4. $S_6 = 0.6703204$; $e^{-0.4} = 0.6703200$;

6. $S_3(0.1) = 0.1-(0.1)^3/3! = 0.0998333$;

   $\sin 0.1 = 0.0998334$; $R_4 \leq 5 \times 10^{-6}$

8. $S_{15}(3) = 0.1411197$; $\sin 3 = 0.1411200$

   $S_3(\pi-3) = 0.1411195$

10. $S_5(0.5) = 0.5210938$; $\sinh 0.5 = 0.5210953$;

    $S_4(0.25) = 1.0314128$; $\cosh 0.25 = 1.0314131$

Problems   P1. $(x+1)e^x = 1 + 2x + 3x^2/2! + 4x^3/3! + \ldots$

              P3. $\sqrt{1 + \sin x} = 1 + x/2 - x^2/2^2 \times 2! - x^3/2^3 \times 3!$

                      $+ x^4/2^4 \times 4! + x^5/2^5 \times 5! - \ldots$

              P10. $B_{10} = 0.0757576$, $B_{12} = -0.2531136$

              P11. $E_6 = -61$, $E_{12} = 2702765$

13

TAYLOR SERIES

INTRODUCTION

After the Fundamental Theorem of the Calculus, Taylor's theorem and
the Taylor series form the most important theoretical and practical
tool of the calculus. They certainly comprise the central concept
of numerical analysis. In the last chapter we developed many familiar functions in series and acquired some facility in their use. We
shall now study some applications of this theory. A first Example
develops a series that approximates the logarithm function at $x = 2$,
even though the function is not defined at all at $x = 0$. Next we
describe Newton's method and give Examples of its use and misuse for
the functions $e^x - 2x - 1$ and $(x-1)/x^2$. Then series integration is
explored with the Sine and Fresnel integrals as Examples. In the
Example of $1/(1-x^2)$ we discuss and then analyze the error in series
integration.

The Exercises provide practice in applying these ideas, includ-
ing studies of the Cosine and Exponential integrals and the error
function. Practice on more difficult applications is given in the

Problem section. In one Problem we consider a new algorithm for finding the zero of a function; it successively finds the zeros of parabolas that just fit the graph of the function. Another Problem describes the theory of convergence for the general iteration function, and another considers the relativistic energy of a moving particle.

TAYLOR'S THEOREM

The remainder term for the series $e^x = 1 + x + x^2/2! + x^3/3! + \ldots$ is $R_n(x) = e^\xi x^{n+1}/(n+1)!$ where ξ is between 0 and x. Since $e^\xi < e^0 + e^x = 1 + e^x$ (do you see why?), which is a number independent of n, $\lim_{n \to \infty} R_n(x) = 0$ for every number x. This means that the infinite series

$$\sum_{i=0}^{\infty} x^i/i!$$

converges to e^x for every x. It is called the Taylor series or expansion of e^x at 0. Every Taylor series (and indeed, every power series) has a constant term $f(0)$ corresponding to the index $i = 0$. We have defined the symbol $\sum_0 b_i$ to be

$$\sum_{i=0}^{\infty} b_i = b_0 + b_1 + \ldots .$$

The Taylor series of a function is thus $f(x) = \sum_0 f^{(i)}(0)x^i/i!$: here we mean that $f^{(i)}(x)$ is the ith derivative of $f(x)$, $f^{(0)}(x) = f(x)$ and that $0! = 1$. It exists for a number x when f has derivatives of all orders on the interval $[0,x]$ and the series converges there. This series amounts to the "polynomial approximation of infinite degree" for f, or, rather, it is the sequence of best polynomial approximations for f. And if the Taylor series for f converges for all x, as it does for e^x, then the sum of the series is $f(x)$ for each x. Nevertheless, in practice we must deal with partial sums $S_n(x)$, and

185

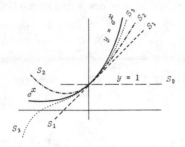

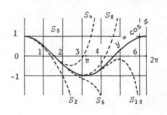

Figure 13.1   Figure 13.2

these polynomials of finite degree n are not *equal* to $f(x)$ unless $f(x)$ is itself a polynomial function. Furthermore, as we can see very clearly in Figures 13.1 and 13.2 for e^x and cos x and we have discovered repeatedly in our calculations, $S_n(x)$ is in general a much better approximation of $f(x)$ when x is near 0 than it is for large x. It is thus desirable in both theory and practice to be able to expand a given function about a point $a \neq 0$ as well as at 0. TAYLOR'S THEOREM *states more generally that the* Taylor series *(or* expansion) *for* f *about* a *is*

$$f(x) = f(a) + f'(a)(x-a) + f''(a)(x-a)^2/2! + \ldots$$

$$= \sum_0 f^{(i)}(a)(x-a)^i/i!$$

with remainder term $R_n(x) = f^{(n+1)}(\xi)(x-a)^{n+1}/(n+1)!$ *for some* ξ *in* [a,x]. Our earlier statement was the special case of this one when $a = 0$; this statement may also be proved from Taylor's theorem at $a = 0$ by applying it to the function $g(x) = f(x+a)$. A Taylor expansion at $a = 0$ is sometimes given the special name of a *Maclaurin series.* (Although Maclaurin did publish this series, he gave Taylor credit for priority. Actually, Gregory and Leibniz knew of Taylor's theorem before Taylor did, and Johann Bernoulli even published something similar in 1694, long before Taylor's announcement in 1712).

186

EXAMPLE: $\ln x$

Consider, for example, the function $\ln x$: it is not defined at all for $x = 0$, so any Taylor expansion for $\ln x$ will have to be at some point $a > 0$. We choose $a = 1$: remember that $\ln x$ has as its sequence of derivatives $1/x$, $- 1/x^2$, $2!/x^3$, $- 3!/x^4$, ..., with $\ln^{(i)}(x) = (-1)^{i-1}(i-1)!/x^i$ and $\ln^{(i)}(1) = (-1)^{i-1}(i-1)!$. Since $(i-1)!/i! = 1/i$, the expansion is

$$\ln x = (x-1) - (x-1)^2/2 + (x-1)^3/3 - \ldots$$

$$\ln x = \sum (-1)^{i-1}(x-1)^i/i.$$

Figure 13.3 depicts the partial sums approximating this series. This is just the series we have already seen for the logarithm function:

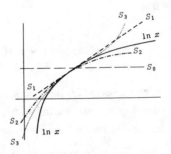

Figure 13.3

if we let $x - 1 = y$ or $1 + y = x$, we may rewrite the above as the familiar series $\ln(1+y) = \sum (-1)^{i-1}y^i/i$. Now imagine that we have used it to calculate $\ln 2$; we may expand $\ln x$ about the point $a = 2$ as

$$\ln x = \ln 2 + (1/2)(x-2) - (1/2)^2(x-2)^2/2 + \ldots$$

$$= \ln 2 + \sum (-1)^{i-1}(x-2)^i/i2^i.$$

187

The remainder term for this series is $R_n(x) = (-1)^n (x-2)^{n+1}/(n+1)\xi^{n+1}$
for some number ξ between 2 and x. To calculate ln 2.1 using this
alternating series, we need merely add up terms until the last one
added is of the magnitude of an acceptable error. If we will accept
four decimal-place accuracy, or an error of 0.00005, the third term
$0.1^3/3X2^3 = 0.0000417$ is an acceptable error. Thus $S_2 + \ln 2 =$
0.7418972 is surely correct to the fourth place (ln 2.1 = 0.7419373
is correct). However, as long as we have calculated the third term,
we may as well add it in to get $S_3 + \ln 2 = 0.7419388$. In fact, if
we add in yet another term, we find that $S_4 + \ln 2$ is correct to
seven decimal places.

NEWTON'S METHOD

Suppose we seek a zero r for a function f. If we have a guess x
that is not far from r, we may expand f in its Taylor series about
x to express $f(r) = 0$ as

$$0 = f(x) + f'(x)(r-x) + f''(\xi)\frac{(r-x)^2}{2!}$$

where the number ξ in the remainder term is somewhere between x and
r. If we then use this Taylor polynomial $S_1(x) = f(x) + f'(x)(r-x)$
to solve backwards for an approximation of r, we get $S_1(x) \doteq 0$ or
$r \doteq x - \frac{f(x)}{f'(x)}$. Why not use this recipe to improve our guess x for
r? Its results will not generally be exactly r, but we may iterate
this process to guess x_0, find $x_1 = x_0 - \frac{f(x_0)}{f'(x_0)}$, then $x_2 =$
$x_1 - \frac{f(x_1)}{f'(x_1)}$, and so on. (This method appeared in Problems P3, P4,
P5 in Ch. 1; P3 in Ch. 2; and P8 in Ch. 4.) How rapidly does this
sequence x_0, x_1, x_2, ... converge to r? The series expression displayed above has its error term built in:

$$r - \left(x - \frac{f(x)}{f'(x)} \right) = \frac{-f''(\xi)(r-x)^2}{f'(x) 2!} \ .$$

Hence, if the inequality

$$\left|\frac{f''(\xi)}{f'(x)}\right| \overset{\leq}{=} M$$

is satisfied for every ξ between r and x, we see that our estimate

$x - \dfrac{f(x)}{f'(x)}$ is in error by less than $\dfrac{M}{2}(r-x)^2$. This estimate doesn't

tell us much if x is a poor guess for r or if M is very large. But suppose that in the case of a given function f we have $M = 2$ for a guess x_0 that is correct in its first decimal place, so we take $(r-x_0) = 0.05$. Then our next guess x_1 has error less than $(0.05)^2 = 0.0025$. Thus x_1 is correct in the second decimal place; x_2 will have an error less than $(0.0025)^2 = 0.0000063$ to be correct in at least four places, and x_3 will be correct to ten places. As a rule of thumb for Newton's method, then, we may expect the number of correct decimal places to be doubled for each iteration (provided $f'(r) \neq 0$ and $f''(r)$ is not enormous).

EXAMPLE: $2x + 1 = e^x$

To illustrate this, we solve the equation $2x + 1 = e^x$ (see Figure 13.4). We apply Newton's method to $f(x) = e^x - 2x - 1$ with $f'(x) \doteq e^x - 2$ and $f''(x) = e^x$. Our recipe is

$$x_{i+1} = x_i - \frac{e^{x_i} - 2x_i - 1}{e^{x_i} - 2},$$

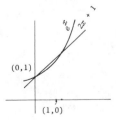

(0,1)

(1,0)

Figure 13.4

and after an inspection of Figure 13.4 we cleverly guess $x_0 = 1\frac{1}{4}$ to calculate $x_1 = 1.2564797$ and $x_2 = 1.2564312$. This second iteration is correct, as we can see by finding $x_3 = x_2$.

Thus our starting guess was correct in the first two decimal places, x_1 in four places, and x_2 in seven (in fact, eight).

As a comparison, we attempt a solution by the method of successive substitutions. The algorithm is x_{i+1} = ln $(2x_i+1)$. Using the same starting guess x_0 = 1¼ we find that it requires ten iterations, x_{10} = 1.2564081, to achieve four-place accuracy! As a check on its startling efficiency, we calculate the error bound for Newton's method. As an estimate for $\left|\dfrac{f''(\xi)}{f'(x)}\right|$ we shall merely compute $\left|\dfrac{f''(r)}{f'(r)}\right|$, since these derivatives are continuous at r:

$$\frac{e^r - 2}{e^r} = 0.4306637,$$

so M = ½ will do, $M/2$ = ¼ and the error behaves like this: $|x_{i+1} - r| < \tfrac{1}{4} |x_i - r|^2$. An algorithm that converges in this fashion, so that each error is a fixed multiple of the square of the preceding error, is said to be *second-order* or *quadratic*. Thus Newton's method is second-order if $M < \infty$ (it is first order if $f'(r)$ = 0 so that r is a *multiple zero*).

To depict the convergence of this method graphically, we magnify the circled region of Figure 13.5 to show in Figure 13.6 the first application of the algorithm to the guess x_0 = 1¼, which yields

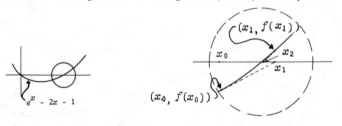

Figure 13.5   Figure 13.6

x_1 = 1.2564797. The line tangent to the graph of $f(x)$ at x_0 is $y(x)$ = $f(x_0)$ + $(x-x_0)f'(x_0)$. If we let x_1 be the point where this line crosses the x-axis, then $y(x_1)$ = 0 = $f(x_0)$ + $(x_1-x_0)f'(x_0)$.

Solving this last equation for x_1 gives the formula for Newton's method, $x_1 = x_0 - \frac{f(x_0)}{f'(x_0)}$. The dotted tangent line depicts this process in the next iteration. Can you imagine from these pictures why this method converges only very slowly when r is a multiple root, so $f'(r) = 0$ as well as $f(r) = 0$?

EXAMPLE: $f(x) = (x-1)/x^2$

The function $f(x) = (x-1)/x^2$ is defined for every argument x except $x = 0$. Its only zero is at $x = 1$. The derivative $f'(x) = (2-x)/x^3$ is zero at $x = 2$ where the tangent line is horizontal. Examine Figure 13.7 to understand this. To apply Newton's method we calculate the algorithm (or "iteration function") to be $\varphi(x) = x - f(x)/f'(x) = (-2x+3)x/(2-x)$.

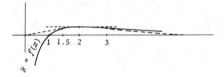

Figure 13.7

A first observation is that $\varphi(2)$ is not defined at all. This corresponds to the fact that the tangent line to the graph of $f(x)$ at $x = 2$ does not cross the x-axis at any point. Of course, Newton's method will work if we start with x_0 close enough to the zero for f at $x = 1$. If $x_0 = 1.1$, then

$$x_1 = 0.9777778$$
$$x_2 = 0.9990338$$
$$x_3 = 0.9999981$$
$$x_4 = 1.$$

A second odd fact about the algorithm $\varphi(x)$ is that if $x_0 = 1.5$, then $x_1 = \varphi(x_0) = 0$ and $x_2 = \varphi(0) = 0$. Here Newton's method has led us to a seeming zero at 0, $x_1 = x_2 = 0$, but the function f is not defined for $x = 0$. In Figure 13.7 you may examine the tangent line to the graph of $f(x)$ at $x = 1.5$ to understand this. Algebraically, φ has two zeros, at $x = 1.5$ and 0.

A third disconcerting fact about the algorithm $\varphi(x)$ is that if we start with $x_0 = 3$, we get

$$x_1 = 9.$$
$$x_2 = 19.285714$$
$$x_3 = 39.687131.$$

Again Figure 13.7 can explain this: the tangent line to the graph of $f(x)$ at $x = 3$ slopes downward away from the zero at $x = 1$. Thus there will be no convergence if x_0 is chosen to be greater than 2. This example illustrates the need for some care in applying Newton's method. However, in this case there is convergence to $x = 1$ if x_0 is chosen between 0 and 1.5; you can see this from the graph of $f(x)$.

EXAMPLE: INTEGRATING THE SINE INTEGRAL WITH SERIES

When we studied integration, we saw that the function $f(x) = x^{-1} \sin x$ has no elementary antiderivative (see Example, Ch. 7; $f(0)$ is defined to be 1, so that f is continuous at 0). Hence the *Sine Integral* $\text{Si}(x) = \int_0^x f(t)dt$, which is useful in the mathematical analysis of wave propogation, had to be evaluated by numerical methods. However, our work with series gives us another handle on that problem. Remember that $\sin x = x - x^3/3! + x^5/5! - \ldots$. Therefore $x^{-1} \sin x = 1 - x^2/3! + x^4/5! - \ldots$, and the integral is

$$\int_0^x \frac{\sin t}{t}\, dt = \int_0^x (1 - \frac{t^2}{3!} + \frac{t^4}{5!} - \ldots)dt.$$

But each term in the series-integrand is quite readily integrated:

$$\int_0^x t^{2n}/(2n+1)!\, dt = t^{2n+1}/(2n+1)(2n+1)! \Big]_0^x = x^{2n+1}/(2n+1)(2n+1)!.$$

192

Since the series for $x^{-1} \sin x$ converges for all x, the series that term-by-term is the integral of the series for $x^{-1} \sin x$ must converge to the value of the definite integral for all x. For all x, then, $Si(x) = x - x^3/3X3! + x^5/5X5! - \ldots$; we have evaluated the integral and, in a sense, found an antiderivative for $x^{-1} \sin x$. Of course, in another sense we have merely exchanged one kind of approximation problem for another. Instead of finding trapezoidal sums for the integral to evaluate $Si(1)$, for instance, we may find partial sums for this series expansion of $Si(x)$ at $x = 1$. Much of the work we do in mathematics looks like that, though; let's investigate the practical value of this new idea. Since this is an alternating series, we may simply calculate terms and sum them until we come to a term whose magnitude is acceptable as an error. In this case, $1/9X9! = 0.0000003$, and the sum of the first five terms $1 - 1/3X3! + 1/5X5! - 1/7X7! + 1/9X9! = 0.9460831$ is correct in all seven decimal places. We found in Chapter 7 that the trapezoidal sum T_{10} for this integral $\int_0^1 t^{-1} \sin t \, dt$ was correct in only three places and was harder to calculate to boot. It is clear that we have made a large stride forward in our computational technique, and also we have acquired a new theoretical tool.

Example: The Fresnel Integral

Another non-elementary integration problem is the *Fresnel integral*:

$$C(x) = \int_0^x \cos\left(\frac{\pi t^2}{2}\right) dt.$$

It is useful in the analysis of diffraction in optics. Since $\cos x = 1 - x^2/2! + x^4/4! - \ldots$,

$$C(x) = \int_0^x \left(1 - \frac{\pi^2 t^4}{2^2 X2!} + \frac{\pi^4 t^8}{2^4 X4!} - \ldots\right) dt$$

$$= x - \frac{\pi^2 x^5}{5X2^2 X2!} + \frac{\pi^4 x^9}{9X2^4 X4!} - \ldots \ .$$

193

To calculate $C(0.3)$, for instance, we need only evaluate this alternating series out to the third term

$$\frac{\pi^4 (0.3)^9}{9 \text{X} 2^4 \text{X} 4!} = 0.0000006$$

and sum to find $C(0.3) = 0.2994010$, which is correct in all seven places. This is a powerful method indeed!

THE ERROR IN SERIES INTEGRATION

Suppose we wish to integrate $f(x) = \sum_0 a_i x^i$ on the interval

$[\alpha, \beta]$: $\int_\alpha^\beta f(x) dx = \sum_0 \frac{a_i}{i+1} (\beta^{i+1} - \alpha^{i+1})$ (assume $0 \leqq \alpha < \beta$).

Now, if in evaluating this definite integral we only use terms of index $i \leq n$ in the integrated series, what is the error? Well, Taylor's theorem says that

$$f(x) = a_0 + a_1 x + \ldots + a_n x^n + f^{(n+1)}(\xi) x^{n+1}/(n+1)!$$

where $0 < \xi < x$. Accordingly, if we replace $f^{(n+1)}(\xi)$ by an upper bound M, $M \geqq |f^{(n+1)}(\xi)|$ for every ξ in $[0, \beta]$, then

$$\left| \int_\alpha^\beta f(x) dx - \int_\alpha^\beta (a_0 + a_1 x + \ldots + a_n x^n) dx \right| \leqq$$

$$\int_\alpha^\beta M \frac{x^{n+1}}{(n+1)!} dx = \frac{M}{(n+2)!} (\beta^{n+2} - \alpha^{n+2}).$$

EXAMPLE: $1/(1-x^2)$

We apply this calculation to an example: $f(x) = \frac{1}{1-x^2} =$

$1 + x^2 + x^4 + \ldots$; this expansion is immediate by substitution of x^2

194

in the geometric series. The remainder term is

$$R_{2n} = \frac{x^{2n+2}}{1-x^2} \; .$$

We could compute $f^{(2n+1)}(\xi)$ from R_{2n}, but it is unnecessary. An upper bound is

$$R_{2n} = \frac{x^{2n+2}}{1-\beta^2}$$

on the interval $[0,\beta]$, so the error in the approximation

$$\int_0^\beta \frac{dx}{1-x^2} \doteq \int_0^\beta (1+x^2+\ldots+x^{2n})\,dx = \beta + \beta^3/3 + \beta^5/5 + \ldots$$

is less than

$$\int_0^\beta \frac{x^{2n+2}}{1-\beta^2}\,dx = \frac{\beta^{2n+3}}{(2n+3)(1-\beta^2)} \; .$$

Since we have an antiderivative in this case, $\int (1-x^2)^{-1}\,dx =$
$\frac{1}{2}\ln\left|\frac{1+x}{1-x}\right|$, we have an approximation when $|\beta| < 1$:

$$\ln\left(\frac{1+\beta}{1-\beta}\right) \doteq 2\beta(1 + \beta^2/3 + \beta^4/5 + \ldots)$$

with error less than $2\beta^{2n+3}/(2n+3)(1-\beta^2)$.

It is simple to check that if $y > 0$ and $\beta = (y-1)/(y+1)$, then $|\beta| < 1$ and $y = (1+\beta)/(1-\beta)$. The above approximation thus gives a rapid method for the calculation of $\ln y$ for every $y > 0$.

EXERCISES

1. .In each of these exercises use Newton's method to solve the indicated equation, beginning with $x_0 = 1$. What is the first value of n for which $x_n = x_{n+1}$? Check your answers.

*a. $2x^2 + 4x - 5 = 0$ c. $\cos x = 0.1$ (so $x = \arccos 0.1$)

 b. $e^x = 4$ (so $x = \ln 4$)   *d. $e^x = \cos x + 1$

2. Apply Taylor's theorem to find the series for $\ln x$ at $a = 7$. Then use your result to calculate $\ln 7.1$ and $\ln 6.1$ if $\ln 7 = 1.9459101$. In each case, choose the appropriate partial sum to achieve five correct decimal places in your final result and give your reasons for that choice.

3. Expand $\sin x$ about $a = \pi/4$, and graph the resulting partial sums S_1, S_2 and S_3 next to the graph of $\sin x$ itself on $[0, \pi/2]$. Compute values at six evenly spaced points. Then on the same graph over the same interval but in another color plot the similar partial sums for the expansion of $\sin x$ about $a = 0$. (Hint: Re-examine Figures 13.1, 13.2, and 13.3.)

4. Find the series for the function $\log_{10} x$ about the point $a = 1$. This function, the *common logarithm* or *logarithm to the base 10*, is the inverse function to $10^y = e^{y \ln 10}$. Hence if $y = \log_{10} x$ then $e^{y \ln 10} = x$ or $y \ln 10 = \ln x$ and $y = \ln x / \ln 10$. Given that $\ln 10 = 2.3025851$, decide how many terms of your series will be required in a partial sum that is correct in five decimal places when $x = 1.11$ and when $x = .5$, and compute these sums. Finally, use these sums to calculate $\log_{10} 1110$ and $\log_{10} 0.0000005$; for each result decide how many decimal places are correct. Does this exercise help to justify the use of the strange number e as a base for natural logarithms?

5. Establish a series representation for the function $\sqrt{x}$ by using a Taylor expansion about $a = 64$. Then use this series to calculate $\sqrt{67.89}$ correct to five places, basing the number of summands you use on an analysis of your remainder term.

6. In our example we found that the partial sum S_4 (2.1) of the expansion of ln x about a = 2 was correct to seven decimal places, and S_2 was correct to four places. How many terms of the expansion of ln x about a = 1 would be required to calculate ln 2.1 correct to four places? Seven places?

*7. Solve $x^5 + 2x^3 + 1 = 0$ by Newton's method.

8. Solve $(e^x-1)^3 = e^{3x} - 3e^{2x} + 3e^x - 1 = 0$ by Newton's method starting with $x_0 = \frac{1}{2}$. Then test a modification of Newton's method for p-fold zeros: if p is the smallest integer for which $f^{(p)}(r) \neq 0$, let $x_{i+1} = x_i - p \dfrac{f(x_i)}{f'(x_i)}$. This modified algorithm may not often

be of much use in practice since one does not generally know p. Nevertheless, in a case of slow convergence it may be worth testing the algorithm for p = 2 and p = 3 for improved convergence. Do you understand the appearance of the factor p in the algorithm in this instance?

*9. Show that the algorithm for Newton's method of finding $\sqrt[n]{a}$ is $x_{i+1} = \dfrac{(n-1)x_i^n + a}{nx_i^{n-1}}$ and use this method to find $\sqrt[3]{7}$ and $\sqrt[3]{19}$. How many iterations were required for each case if x_0 = 2?

*10. The *Cosine Integral* Ci(x) is used in the analysis of wave propagation; it is defined to be

$$Ci(x) = \gamma + \ln x + \int_0^x t^{-1}(\cos t - 1)dt,$$

where γ = 0.5772157 is the Euler number. The negative of the integral itself is called Cin $(x) = \int_0^x t^{-1}(1 - \cos t)dt$; the value of the integrand at t = 0 is taken to be 0 so that it is continuous there. There is no elementary antiderivative for $(1 - \cos t)/t$; hence this integral must be evaluated by numerical methods. Show that Cin $x = \dfrac{x^2}{2 \times 2!} - \dfrac{x^4}{4 \times 4!} + \ldots$ and compute Cin (0.7) correct to six places. (Give the reason you believe that your answer is correct to six places. Compare your results to the four-place accuracy of the trapezoidal sums in Exercise 9, Ch. 7.)

*11. The *Exponential Integral* Ein $(x) = \int_0^x t^{-1}(1 - e^{-t})dt$ (sometimes $E_1(x) = \text{Ein}(x) - \ln x - \gamma$ is called the "exponential integral") is another function that is used in applied mathematics and that is not expressible in terms of elementary functions. Show that Ein $x =$

$x - \dfrac{x^2}{2 \text{X} 2!} + \dfrac{x^3}{3 \text{X} 3!} - \ldots$ and evaluate $\text{Ein}(x)$ correct to six decimal

places. (Give the reason you believe your answer is correct to six places.)

*12. The *error function* or *probability integral* (see also Exercise 8, Ch. 8)

$$\text{erf } x = H(x) = \frac{2}{\sqrt{\pi}} \int_0^x e^{-t^2} \, dt$$

is not expressible in terms of elementary functions. Show that erf $x = \dfrac{2}{\sqrt{\pi}} (x - x^3/3 + x^5/5 \text{X} 2! - x^7/7 \text{X} 3! + \ldots)$ and calculate erf(0.2) correct to six decimal places. Give your reasons for believing that your answer is correct in its sixth place.

*13. Calculate ln 2 using the approximation $\ln\left(\dfrac{1+\beta}{1-\beta}\right) =$ $2\beta(1 + \beta^2/3 + \beta^4/5 + \ldots)$ from our Example. Decide from the error term $2\beta^{2n+3}/(2n+3)(1-\beta^2)$ how many summands are necessary to insure six-place accuracy and use that sum.

14. Use Newton's method to construct a general algorithmic scheme for solving quadratic equations of the form $x^2 - 2\beta x + \gamma = 0$, where $\beta^2 > \gamma$. Then use your algorithm to find solutions for the following equations, starting with $x_0 = \beta - 1$ and $x_0 = \beta + 1$:

$$x^2 - 5x + 6 = 0$$
$$x^2 + 2\pi x - e = 0.$$

198

Finally, discuss the equation $x^2 - 2\beta x + \gamma = 0$ in case $\beta^2 = \gamma$ or $\beta^2 < \gamma$. What happens to your algorithm in these cases?

PROBLEMS

P1. Use Taylor's theorem to derive a higher-order approximation method, similar to Newton's method, for finding the zeros of a function $f(x)$. Show that if one uses the Taylor expansion of $f(r)$ at an approximation x_i of a root r of $f(x) = 0$ and neglects terms of order greater than two, that

$$x_{i+1} = x_i - \frac{f'(x_i) - \{f'(x_i)^2 - 2f(x_i)f''(x_i)\}^{\frac{1}{2}}}{f''(x_i)}$$

is a better approximation than x_i. (The choice of the negative sign in the numerator is made to minimize the numerator.) This formula fits a parabola tangent to the graph of f at x_i and solves for one of its zeros. It is called *Cauchy's method*.

Compare the rate of convergence of this method with that of Newton's method and the method of successive substitutions for our Example function $f(x) = e^x - 2x - 1$. Discuss the advantages and disadvantages of this higher-order method.

P2. Integrate $\int_0^1 \frac{e^x - 1}{x}\, dx$ by means of a Maclaurin series expansion of the integrand. Analyze the error and decide how many terms will be necessary to have the series approximation to this integral accurate to six places. Then compute that sum.

P3. Find $\int_0^{\pi/4} \frac{1 - \cos x}{\sqrt{x}}\, dx$ accurately to six places. Do this by series integration. Give your reason for believing that your answer is indeed correct to six decimal places.

P4. Find $\int_0^1 e^{x^5}\, dx$ accurately to six places by means of series integration. (Remember, e^{x^5} means $e^{(x^5)}$, not $(e^x)^5$.) Give an argument based on the error term for your particular choice of a partial sum.

P5. Derive the Taylor series for sin x expanded about the point $a = \pi/4$ as in Exercise 3. Then use this series to derive a series for $\int_{\pi/4}^{\pi/3} \sin x^2 \, dx$. Analyze the error term for this series to decide how many terms are necessary for a partial sum to be correct to six decimal places and form that sum.

P6. In Newton's method, and also in the method of successive substitutions, a solution is found to an equation $f(x) = 0$. The solution s is the limit of a sequence x_0, x_1, x_2, ... defined by an *iteration function* φ, so that $x_{i+1} = \varphi(x_i)$ for each integer i, once the initial estimate x_0 is chosen. Suppose the iteration function φ has derivatives of all orders at s and consider its Taylor expansion at s (compare Problem P3, Ch. 5): $\varphi(x_i) =$

$$\varphi(s) + (x_i - s)\varphi'(s) + (x_i - s)^2 \varphi''(s)/2! + \ldots .$$

Let the nth error be defined as $\varepsilon_n = x_n - s$: show that

$\lim_{n \to \infty} \varepsilon_{n+1}/\varepsilon_n = \varphi'(s)$. Next assume that $|\varphi'(s)| < 1$ and prove that the sequence of iterants does indeed converge to s. Conversely, argue that if $|\varphi'(s)| > 1$, then the sequence x_0, x_1, ... cannot converge to s.

Given the problem of finding a root s for $f(x) = 0$, suppose that the iteration function φ, which you construct for successive substitutions, has a derivative and $|\varphi'(s)| > 1$, so that it does not converge. Describe another substitution algorithm for the same problem that is guaranteed to converge.

Show that $\varphi'(s) = 0$ whenever φ is an iteration function given by Newton's method and $f'(s) \neq 0$. Explain why the speed of convergence is called "quadratic" in this case.

P7. The relativistic energy of a moving particle is given by $E = m/\sqrt{1 - \beta^2}$ where m is the mass and β is the particle speed expressed as a fraction of the speed of light. When $\beta = 0$, the particle has "rest mass" m. Expand $E(\beta)$ in a Taylor series. Then argue that if β is small, say the speed is less than 300 km/sec (which is one one-thousandth that of light), then E is well-approximated by the sum of the rest mass and the classical kinetic energy $m\,\beta^2/2$.

Answers to Starred Exercises

Exercises   1a. $x_3 = x_4 = 0.8708287$

               1d. $x_4 = x_5 = 0.6013468$

                7. -0.7331569

                9. $x_3 = 1.9129312 = \sqrt[3]{7}$

                        $x_4 = 1.8019831 = \sqrt[5]{19}$

             10. $\mathrm{Cin}(0.7) = 0.1200260$

             11. $\mathrm{Ein}(1/\pi) = 0.2946699$

             12. $\mathrm{erf}(0.2) = 0.2227026$

             13. $S_5 = 0.6931471$

14

DIFFERENTIAL EQUATIONS

INTRODUCTION

Applications of the calculus depend on interpretations of the derivative, such as the slope of a graph, a velocity or an acceleration, marginal profit or cost or revenue, a rate of growth or of decay. For example, acceleration is the derivative of speed for a moving vehicle. Thus if the acceleration of an object is known to be constantly 7, then its speed $s(t)$ as a function of time satisfies the equation $s'(t) = 7$. This is called a differential equation: it is an equation involving the derivative of a function. The solution to this equation is not a number; it is the function $s(t) = 7t + C$, where 7 is the constant acceleration and C is the number $s(0)$, the value of the speed at the time coordinate 0. In general, in differential equations the unknowns do not stand for numbers but for functions, and the solutions are functions.

In real-life applications where we wish to know a function, we frequently are able to understand the relationship(s) that the derivative(s) of a function must satisfy. Hence differential equations constitute the most important way in which the calculus is used to solve practical problems.

This chapter begins with the Example of exponential growth and some definitions. We then discuss the case of separable variables and illustrate it with the Example of the spread of rumors. Next, two series methods of solving differential equations are described, and Examples are given. Then an Example of a stepwise process shows that, with a constant amount of arithmetic, accuracy may suffer when we subdivide the interval over which solutions are calculated. Exercises offer practice in solving first order differential equations by series and by separation of variables. (This chapter may be regarded as being primarily concerned with series.)

In the Problems we define and discuss simultaneous sets of differential equations, second order equations, Euler's method, and the Heun method. Series solutions that are expanded about some point $x \neq 0$ provide another Problem as does a second order equation whose leading coefficient vanishes at the initial point.

EXAMPLE: $y' = ky$ AND EXPONENTIAL GROWTH

In Chapter 8 we discussed the problem of radiocarbon dating. We knew that the radioactive atoms of C^{14} in an ancient axe handle had been decomposing at a rate proportional to the amount of C^{14} present. That is, if $y(t)$ is the amount of C^{14} present in the axe handle at time t years after the wood ceased to grow, then there is a constant k of proportionality such that $y'(t) = k\, y(t)$. We solved this problem by guesswork then. Here is a more systematic method of attack: the equation may be written as $y'(t)/y(t) = k$. We know that

$y'(t)/y(t) = \frac{d}{dt} |\ln y(t)|$ is a derivative. By the Fundamental Theorem of the Calculus,

$$\ln|y(t)| = \int_0^t \frac{y'(x)}{y(x)}\, dx = \int_0^t k\, dx = kt.$$

Thus $|y(t)| = e^{kt}$.

Since our choice above of 0 for the lower limit of integration was arbitrary, we may add any constant c to kt to find another function $kt + c$ that also satisfies the requirement on $\ln|y(t)|$. This defines a family of functions $|y(t)| = e^{ky+c} = e^c e^{kt}$. Now e^c may be any positive number, and $-y(t)$ is a solution to our problem whenever $y(t)$ solves it. Thus we may describe the family of solutions as $y(t) = Ce^{kt}$ for various constants C. Here C may be either positive or negative; the constant function $y(t) = 0$ is also a solution. Figure 14.1 graphs this family of functions; there is exactly one

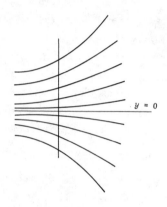

Figure 14.1

graph of such a function going through a given point (x,y) of the plane. Hence we could single out a specific solution $y_0(t)$ by specifying that its graph go through $(0,3)$, for instance, which makes $Ce^0 = 3$ or $C = 3$. Thus $y_0(t) = 3e^{kt}$.

SOME DEFINITIONS

The above equation, $y'(t) = k\,y(t)$, is a *differential equation*. In Chapter 8 we saw similar differential equations, which described the growth of money with compound interest and the cooling of a hot body. In general, a differential equation, which we often call a *DE* for short, is an equation involving a function $y(t)$, the variable t, and the derivative function $y'(t)$. Sometimes a DE will involve the second or subsequent derivatives $y''(t)$, $y'''(t)$, ...; these are called

204

second order DEs, third order DEs, ... We shall not consider
higher order differential equations here (see Problems P2 and P8).
An equation that involves the first derivative $y'(t)$, plus $y(t)$ and
x perhaps, is called a *first order DE* (they are sometimes called
ordinary DEs to distinguish them from equations involving partial
derivatives).

Every integration problem may be considered to be a first order
differential equation: to find an indefinite integral $\int f(x)\ dx$ is
to find a solution to $y'(x) = f(x)$. Similarly, to find $\int_a^b f(x)\ dx$
is to find a particular solution to $y'(x) = f(x)$ for which $y(a) = 0$,
and then the definite integral is $y(b)$ (can you see why?). This fact
can give us humility: we know now that we cannot solve every DE of
the form $y'(x) = f(x)$, let alone those DEs that involve $y(x)$ as well.
Examples we have seen of integration problems for which no elementary solution exists include the definitions of Bessel functions;
elliptic integrals; the error function; the Sine, Cosine, and Exponential functions; and others. But there are many DEs that may be
solved. Unfortunately there are many methods for solving them, and,
just as for finding antiderivatives, no one of these methods is sure
to work on a given problem.

Separable Variables

One method that works on many simple DEs that arise in applications
is the *method of separation of variables*. Our axe handle problem
above is an example for this method. The DE is $y'(t) = k\ y(t)$,
which we shall write more simply as $y' = k\ y$. To solve it, rewrite
it with all the symbols that involve y appearing in a factor multiplying y' on the left-hand side, so that the right-hand side of the
equation is a function of x only, with no appearance of the symbol y.
Rewritten, the equation is $\frac{1}{y} y' = k$. The resulting equation is now
in principle the equality of two derivatives, each with respect to x.
The antiderivatives ($\ln |y|$ and kx) of each side may now be found.
The Fundamental Theorem says that two functions with the same derivative must differ by a constant. For each choice C of a constant,

then, $\ln |y| = kx + C$ is a solution to this DE.

In general, the trick is to arrange the DE in the form $F(y)y' = G(x)$. This is not always possible; when it is, the DE is said to have *separable variables*. Then the solution is of the form

$$\int F(y) \; y' \; dx = \int G(x) \; dx.$$

(Here the indefinite integral $\int G(x) \; dx$ stands for the family of antiderivatives, of functions whose derivative is $G(x)$.)

EXAMPLE: THE RUMOR DE

For a simple model of the spread of a rumor, suppose that each person who has heard it will meet 7 people per day and tell them all. Some of them will have already heard it. If we let $H(t)$ be the number of people who have heard the rumor at time t, out of a total

population P, then each day each of them will inform $7(1 - H(t)/P)$ persons who have not yet heard it. This gives the DE

$$H' = \frac{7}{P} H(P\text{-}H).$$

The variables are separable. The rewritten equation is

$$\frac{P}{H(P\text{-}H)} H' = 7.$$

The solution will be

$$\int \frac{P}{H(P-H)} H'\, dt = \int 7\, dt = 7t + C.$$

To integrate the left-hand side, notice that

$$\frac{P}{H(P-H)} = \frac{1}{H} + \frac{1}{P-H}$$

and

$$\int \frac{P}{H(P-H)} H'\, dt = \int \frac{H'}{H}\, dt + \int \frac{H'}{P-H}\, dt = \ln |H| - \ln |P-H| + C.$$

(If this is at all confusing, replace $H'\, dt$ by dH in these integration problems.) Since $0 \le H \le P$, the absolute value symbols may be deleted. This yields

$$\ln H - \ln(P-H) = 7t + C$$

$$\ln\left(\frac{H}{P-H}\right) = 7t + C$$

$$\frac{H}{P-H} = e^{7t + C} = k\, e^{7t}, \; k > 0.$$

Solving for H as a function of t gives

$$H = k\, e^{7t}(P-H)$$

$$H(1+ke^7) = k\, P\, e^{7t}$$

$$H(t) = \frac{k\, P\, e^{7t}}{1+ke^{7t}} = \frac{k\, P}{k+e^{-7t}}.$$

Here P is the total population, t is the time in days, and k is a constant determined by the initial value of $H(t)$ at $t = 0$.

EXAMPLE: SERIES SOLUTION BY COMPUTED COEFFICIENTS FOR $y' = 2xy$

Perhaps the most widely applicable methods of solving DEs are the two methods of finding the Taylor series of a solution. These methods will often provide a theoretical solution, even in closed form, and also they yield numerical solutions for given initial values.

As an example, we solve $y'(x) = 2xy(x)$, this DE would usually be written simply as $y' = 2xy$. We wish to find the set of all functions $y(x)$ that satisfy this DE. In fact, this DE has separable variables. Hence the method described above will work. We see that $2x = y'/y = \frac{d}{dx} \ln |y|$, so $\ln |y| = \int 2x\,dx = x^2 + c$, and thus $y = Ce^{x^2}$. That is, the set of solutions for $y' = 2xy$ includes all functions of the form $y(x) = Ce^{x^2}$ for some real number C (see Figure 14.2).

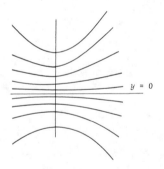

$y = 0$

Figure 14.2

We put this knowledge aside, however, and use the *method of computed coefficients* to find a Taylor expansion for $y(x)$ about the point $a = 0$. The value of y at $x = 0$ is arbitrary (that is, it may be any real number); we denote it by $y(0) = C$ and compute the successive derivatives of $y(x)$, first the functions and then their values at $x = 0$. All this can be done from the information we are given,

208

namely, $y' = 2xy$ and $x = 0$ and $y(0) = C$. The computation is compiled in Table 14.1.

TABLE 14.1

Value at x	Value at $x = 0$
$y = y(x)$	C
$y' = 2xy$	0
$y'' = 2xy' + 2y$	$2C$
$y''' = 2xy'' + 4y'$	0
$y^{(4)} = 2xy''' + 6y''$	$12C$
$y^{(5)} = 2xy^{(4)} + 8y'''$	0
$y^{(6)} = 2xy^{(5)} + 10y^{(4)}$	$120C$

.

.

.

$$y^{(n)} = 2xy^{(n-1)} + (2n-2)y^{(n-2)} \quad (2n-2)y^{(n-2)}(0)$$

The Taylor series for the solution function is thus

$$y(x) = C(1 + x^2 + x^4/2 + x^6/6 + \ldots).$$

In this case, the series is readily recognized as that for e^{x^2}. To check this, we try $y(x) = Ce^{x^2}$ in the DE and see that it is satisfied.

If, however, we did not recognize the series to be that of some familiar function, in closed form, we nevertheless would have a series representation of a family of solutions. Furthermore, suppose that we are given a specific number for the *initial value* $C = y(0)$ as part of the problem we are to solve. Then we can calculate the value of a specific solution function y at some other point x by means of this series. The remainder term for the partial sum S_n of this series was shown in Chapter 13 to be less than $Ce^{x^2}x^{n+1}/(n+1)!$. Here we may roughly estimate e^{x^2} as less than 3^{x^2}. If $x = 1$, for example, and C is given as $\frac{1}{2}$, then six-place accuracy may be achieved with S_9. Suppose though that we wish to calculate $y(2)$: to have six decimal places correct requires that we evaluate S_{19}, which is a lot of summing.

EXAMPLE: SERIES SOLUTION BY UNDETERMINED COEFFICIENTS FOR $y' = x - y$

Instead of calculating the coefficients of the Taylor series for $y(x)$ directly from the initial value $x = 0$ and $y(0) = C$, we may proceed algebraically via the *method of undetermined coefficients*. We illustrate this alternative scheme with a new DE for an example: $y' = x - y$.

To begin, suppose that y does have a Taylor series expansion within some positive radius of $a = 0$: $y(x) = a_0 + a_1x + a_2x^2/2! + \dots$.
Then $y'(x) = a_1 + a_2x + a_3x^2/2! + \dots$, and $x - y =$
$-a_0 + (1-a_1)x - a_2x^2/2! - a_3x^3/3! + \dots$. Since $y' = x - y$, the coefficients of like powers of x must be equal, since a power series expansion for a function is unique (see Chapter 12). Thus the constant terms a_1 for y' and $-a_0$ for $x - y$ must be equal: $a_1 = -a_0$.
Similarly $a_2 = 1 - a_1 = 1 + a_0$, and then $a_3 = -a_2$, $a_4 = -a_3$, and $a_i = -a_{i-1}$ for $i \geq 3$. Thus when a_0 is determined, then every coefficient of the Taylor series $\sum_0 a_i x^i/i!$ for $y(x)$ is completely specified:

$$y(x) = a_0 - a_0x + (1+a_0)x^2/2! - (1+a_0)x^3/3! + (1+a_0)x^4/4! - \dots .$$

This series is almost recognizable; aside from the first two terms it is the series $(1+a_0)\left(\sum_0 (-x)^i/i!\right) =$

$1 + a_0 - (1+a_0)x + (1+a_0)x^2/2! - (1+a_0)x^3/3! + \dots$, which is the Taylor series for the function $(1+a_0)e^{-x}$. Therefore we may simply adjust the first two terms of this latter series to agree with that for y, which is seen to be $y(x) = (1+a_0)e^{-x} - 1 + x$. This constitutes a set of solutions for the DE $y' = x - y$ where there is a

210

different solution for each differing initial value $y(0) = a_0$. This
fact corresponds in Figure 14.3 to the fact that exactly one curve

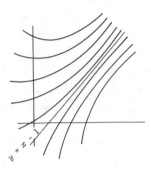

Figure 14.3

goes through each point $(x, y(x))$ of the plane. The solution $y =$
$Ce^{-x} - 1 + x$ is in closed form; however, the series solution $y(x) =$
$a_0 - a_0 x + (1+a_0)(x^2/2! - x^3/3! + ...)$ is a satisfactory theoretical
solution to our differential problem. As an example, assume that
$y(0) = 1$ (so the solution function is $y(x) = 2e^{-x} - 1 + x$); we shall
calculate $y(0.5)$ from the series. The error for this alternating
series will be less than the first term omitted from a partial sum,
so for five-place accuracy we sum $S_6 = 0.7130642$. The first omitted
term is $-2(0.5)^7/7! = -0.0000031$, and the theoretical value is
0.7130613.

(This method, of undetermined coefficients, gives us insight
into DE's in general. Since in principle it must always work, it
shows us intuitively that if the coefficients of a DE are functions
that themselves have power series expansions, then a series solution
for that DE always exists. It also may be easy to see a regularity
in the determination of these coefficients and thus to guess the general term of the series for y. Nevertheless, the applications of
this method often lead into complicated arithmetic. The first method
we saw, of computed coefficients, may, then, be the simpler one to
use.)

Example: A Stepwise Process

Could we improve our accuracy in the last calculation above by breaking up the interval [0, 0.5] into, say, two pieces [0, 0.25] and [0.25, 0.5] and using our solution above at 0, $y(0) = 1$ to establish a new solution at 0.25? That is, suppose that we use the partial sum $S_3(0.25) = 0.8072917$ to approximate $y(0.25)$, then use this value of y together with a series expressing $y(0.5)$ as an expansion at $a = 0.25$ to calculate a new partial sum for $y(0.5)$. Would this two-step procedure yield more accuracy in our estimate for $y(0.5)$? Such increased accuracy might be expected; using smaller steps in integration processes certainly improves the accuracy, for instance. At $x = 0.25$ we have

$$x = 0.25$$
$$y \doteq 0.8072917$$
$$y' = x - y \doteq -0.5572917$$
$$y'' = 1 - y' \doteq 1.5572917$$
$$y''' = -y''' = -1.5572917$$

and the series is

$$y(x+h) = y(x) + h\,y'(x) + h^2 y''(x)/2! + h^3 y'''(x)/3! + \ldots$$

where both x and h are equal to 0.25. This sum is $y(0.5) = 0.7125787$, which is only correct to two places. Our hopes are thus dashed. By summing the first four terms of each of two series, and also calculating the coefficients in a second series, we got a much worse result than we originally had with the sum of seven terms of one series (see Figure 14.4). Of course, if we had summed seven terms of the series for each of our two steps, we *would* have achieved increased accuracy (see Problem P5).

We may understand this fact to suggest that the way a DE guides us to numerical answers is basically divergent. That is, a small

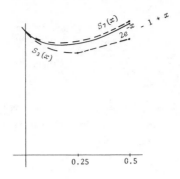

Figure 14.4

error in an earlier step is amplified each time it is used in a later calculation. This contrasts with Newton's method for finding zeros, for instance, where an error in an intermediate step will be corrected in later stages.

In practice, DE's may themselves be defined numerically so that, for example, one is posed the problem $y' = f(x,y)$ where the function f is not known in closed form. (In applications, the value of f may be read on a meter.) Then it is numerically unsound to calculate y'', y''' and so forth as limits of difference quotients. In such cases, the breaking down of the interval into steps may be the only practical method available (see Problem P3 and P4). We have included the above example, though, as a warning that the stepwise procedures that are universally used with large computers may often be inappropiate for hand-held calculators.

Exercises

1. Treat each of the following easy integration problems as a DE and solve it twice: (1) by the method of computed coefficients and (2) by the method of undetermined coefficients. Find families of solutions that are series expansions about the point 0.

a. $y' - 3x + 4 = 0$ c. $y' - xe^{x} = 0$

b. $y' + e^{x} = 0$ d. $y' + k\,y = 0$

213

*2. Calculate $S_9(1)$ and $S_{19}(2)$ for the series $y(x)$ = $\frac{1}{2}(1 + x^2 + x^4/2! + x^6/3! + \ldots)$ of our example of the method of computed coefficients.

3. Use the method of undetermined coefficients to solve the DE $y' = 2xy$, which was solved in the Example of the method of computed coefficients.

4. Use the method of computed coefficients to solve the DE $y' = x - y$, which was solved in the Example of the method of undetermined coefficients.

*5. Solve $xy' = 2y$ by series. Give the solution $y_0(x)$ for which $y_0(1) = 2$ and also describe a family of solutions, one solution for each starting value $C = y(1)$. (Check your answer.)

*6. Solve $y' = x + 2xy$ by series. Give a family of solutions, one for each starting value $a_0 = y(0)$. Then compute the value $y_0(0.7)$ correct to six places for the solution $y_0(x)$ for which $y_0(0) = 0$. Finally, identify the series solution you have found with an elementary function in closed form and compute $y_0(0.7)$. (Check your answers.)

7. Solve $y' = y^2 + x$ by series (use computed coefficients) when the initial value of the solution is $y(0) = 0$. Then estimate $y(0.5)$ correct to six decimal places and offer a plausible argument that you have considered enough terms in your partial sum to achieve that accuracy.

*8. Solve $y' = x^2 + y$ by series. Find a family of solutions, one for each initial value $y(0) = C$. Then evaluate the solution $y_0(1)$ correct in six decimal places if $y_0(0) = 0$. Give your reason for believing that your sum is indeed correct to six places.

9. Solve $y' = (x^2-y)/(1-x)$ by series in case $y(0) = 0$. Then calculate $y(0.1)$ correct to six places and give your reason for believing that you have summed enough terms to be correct in the sixth place.

10. Suppose a bacterial culture has a tendency to exponential growth; that is, a tendency to satisfy $y' = ky$. Suppose also that the food

supply is limited and will only support a total population A. The growth rate itself must then decrease as the population increases. Assume that the growth rate y'/y decreases exponentially according to the rule

$$y' = k\, e^{-Bt}\, y, \quad B > 0.$$

This equation has its variables separable. Integrate it to get *Gompertz's growth curve*.

Let t be measured in hours and suppose an initial growth rate of 9% per hour for an initial population count of 100 per cc. Let the food supply put a limiting upper bound A on the population of 10^4 per cc. Find the population counts at the end of 2 days and 10 days. (Hint: The upper bound A is the limit of the values of your function $y(t)$ as t goes toward infinity.)

11. Make a mathematical model for the spread of an epidemic. Let $I(t)$ be the number of people infected at time t in days out of a total S of suspectible people. Suppose that the number who catch the

disease each day is a constant multiple of the number who have it times the number who are suspectible but not yet infected. Write down the appropriate DE, separate the variables, and integrate it. Then calculate the values of I after two weeks, two months, and a year for the following case. Three people are infected among a susceptible population of 1000 at time $t = 0$, and the Public Health Service counts 96 persons infected on the seventh day.

215

12. Make a graph of the solution function for the rumor DE $H(t) = k \, P/(k + e^{-7t})$. Assume that $P = 10^6$ and that one person begins the rumor when $t = 0$.

13. Suppose that brine containing 3.5 kilograms of salt per 100 liters is flowing into a 10,370 liter tank. Let the tank start out full of fresh water and be stirred so as to be perfectly mixed at

all times. If the brine flows in at the rate of 57 liters per minute and the tank overflows at the same rate, how many kg of salt are in the tank after 5 minutes, 5 hours, 5 days, 5 weeks? (Hint: Write a DE for the amount $S(t)$ of salt in the tank at time t and separate the variables.)

PROBLEMS

P1. Simultaneous sets of DEs may be solved by series methods in quite similar fashion to the examples we have seen. Suppose there are two functions $x(t)$ and $y(t)$ and it is known that $x(0) = 1$ and $y(0) = 1$ and that the derivatives satisfy $x' = 2x + y$ and $y' = y - x$. Assume there is a series $x(t) = \sum_0 a_i x^i$ and also a series $y(t) = \sum_0 b_i x^i$ and solve these DEs simultaneously to determine the coefficients $a_0, b_0, a_1, b_1, \ldots$. Then discuss the remainder terms for these two series and give $x(0.3)$ and $y(0.3)$ correct to six decimal places.

P2. The DE $y'' = 2y + 3x$ is an example of a *second order differential equation*. The method of undetermined coefficients will work on it as well as it does for first order DEs. Find the Taylor series for a family of solutions $y(x)$. The functions in your family should depend on two unspecified constants so that there will be a solution determined whenever initial values are given to both $y(x)$ and $y'(x)$. Give the solution in case $y(0) = y'(0) = 1$ and compute $y(1)$ accurately to six places.

216

P3. *Euler's method* for the numerical solution of DEs may be illustrated in our example $y' = 2xy$: to compute $y(1)$ when $y(0)$ is known to be $1/2$, we divide the interval $[0,1]$ into n subintervals, which in this problem will have equal length $h = 1/n$. Let $y_0 = y_1 = y(0)$, $y_2 = y_1 + 2y_1/n^2$, and in general define $x_i = i/n$ and $y_i = y_{i-1} + 2x_{i-1}y_{i-1}/n$. Then y_n is the approximation we seek for $y(1)$. Calculate y_n for $n = 5$ and 10 for the above problem. Then draw a graph that depicts your calculation for the case $n = 5$. Observe the poor accuracy of this method; it can be shown that the error is roughly proportional to h. Assuming that, how large must n be in this problem for six-place accuracy?

P4. The *modified Euler method* or *Heun method* is an example of a *predictor-corrector* method. In Problem P3 above, the estimate y_i in the Euler method is predicted by the tangent, with slope $2x_{i-1}y_{i-1}$, to the solution curve at (x_{i-1},y_{i-1}). In the Heun method the "predicted" y_i is used to construct a "corrected" linear approximation with slope the average of y' computed at two points, rather than the slope of the tangent line at y_{i-1}. To solve the same DE, $y' = 2xy$ with $y(0) = \frac{1}{2}$, we begin by predicting $y_1 = y(0) + y'(0)h = \frac{1}{2}$. The slope at (x_1,y_1) is $2x_1y_1 = 1/n$. Define $\overline{y}_1 = y(0) + \frac{h}{2}(0 + 1/n) = \frac{1}{2} + \frac{1}{2n^2}$; continue with $y_2 = \overline{y}_1 + 2x_1\overline{y}_1h$ and

$$\overline{y}_2 = \overline{y}_1 + 2x_1\left(\frac{y_1+y_2}{2}\right)h.$$

In general, $y_i = \overline{y}_{i-1} + 2x_{i-1}\overline{y}_{i-1}h$ and

$$\overline{y}_i = y_{i-1} + 2x_{i-1}\left(\frac{\overline{y}_{i-1}+y_i}{2}\right)h.$$

Calculate $\overline{y}_n$ for $n = 5$ and 10. Compare your results to those of Problem P3.

*P5. Calculate the value $y(0.5)$ for the solution $y(x)$ to the DE $y' = x - y$ for which $y(0) = 1$. Do this calculation by first summing

$S_6(0.25)$ for the Taylor series given in the example: $y(x) =$ $1 - x + 2(x^2/2! - x^3/3! + \ldots)$. Then use $S_6(0.25)$ as an approximation of $y(0.25)$ to calculate the coefficients for the Taylor series for this same function $y(x)$ expanded about $a = 0.25$. Finally evaluate the partial sum $S_6(0.25)$ of this new series to estimate $y(0.5)$. Compare your answer with the single-step result and also with the two-step result from our example. How many terms of the series expansion about $a = 0$ would be required to achieve the same accuracy in a single step?

*P6. Since some functions, like $\ln x$ or $\frac{1}{x}$, have no Maclaurin series (at $a = 0$), we cannot hope to be able every time to solve a given DE by seeking coefficients for a Maclaurin series. But the methods we have developed may prove effective in finding an expansion of the solution about some other point, say $a = 1$. Show that the DE $y' = (x+y)/x$ is an example of these remarks: its solution cannot be represented as a power series in x. Then solve it by determining the coefficients of a solution expressed as a power series in $x - 1$. Then calculate $y(1.5)$ for the solution $y(x)$ for which $y(1) = 2$; do this by summing enough of your series to guarantee six-place accuracy. Finally, identify your series in closed form as an elementary function and check it.

*P7. Solve $y' = y - 2 \cos x$ when $y(0) = 1$. Then evaluate $y(2)$ correct to six places, using the remainder term to decide how many terms to sum. Finally identify your solution in closed form and check it in the DE. Can you now describe the family of solutions, one solution for each real initial value?

P8. Suppose we must solve the second order DE $2xy'' + y' + y$ with the initial value $y_0(0) = 0$. Use the method of undetermined coefficients to find a family of solutions to this DE, yet show that your family contains no nontrivial solution $y_0(x)$. This is related to the fact that the coefficient of y'' vanishes at $x = 0$. Try again for solutions using undetermined coefficients for the series $y(x) =$ $\sum_0 a_i x^{\alpha+i}$, where α is a real number. Show that there is a value

$\alpha_0 \neq 0$ for which this series gives a nontrivial solution for $y_0(x)$. Finally calculate the value $y_0(0.2)$ correct to six places.

P9. Describe at least one plausible situation in a field of your own current interest, perhaps biology or business or chemistry, where differential equations may be applied to obtain a useful numerical solution. Discover such a real-life situation by surveying a current issue of an appropriate journal in your field. (See the Bibliography for some suggested journal titles.)

Answers to Starred Exercises and Problems

Exercises
2. $S_9(1) = 1.3591408$, $S_{19}(2) = 27.299075$

5. $y_0(x) = 2x^2$, $y(x) = Cx^2$

6. $y(x) = a_0 + (a_0 + \frac{1}{2})(x^2 + x^4/2! + x^6/3! + \ldots) =$

$- \frac{1}{2} + (a_0 + \frac{1}{2})e^{x^2}$

$y_0 = - \frac{1}{2}(x^2 + x^4/2! + x^6/3! + \ldots)$

$S_8 = -0.8161581 = - \frac{1}{2}e^{(0.7)^2}$

8. $y(x) = C + Cx + Cx^2/2 + (C/6 + 1/3)x^3 +$

$\ldots + (C+2)x^n/n! + \ldots$

$y_0(1) \doteq S_{11}(1) = 0.4365637$

Problems
P5. $y(0.5) \doteq 0.7130614$

P6. $y(x) = x(C + \ln x)$

P7. $y(x) = 1 - x - x^2/2 + x^3/3! + x^4/4! - x^5/5! - \ldots$

$= \cos x - \sin x$

In general $y(x) = \cos x - \sin x + Ce^x$.

APPENDIX: SOME CALCULATION TECHNIQUES AND MACHINE TRICKS

Introduction

This appendix offers some suggestions that will make your work with your calculator faster and more efficient. These suggestions cover "invisible registers" and program records, the rewriting of formulas and planning of a calculation, constant arithmetic, factoring integers and finding integer parts, synthetic division and the evaluation of polynomials or series, and "artificial" scientific notation. Also a method is given for converting decimal yards or hours to yards-feet-inches or hours-minutes-seconds.

Next there is a discussion of roundoff, overflow, and underflow; followed by a method for handling large exponents. The appendix closes with a few facts about the machines themselves and how to avoid damaging them, plus some references for those who wish to read further about them.

Invisible Registers

Registers are the electronic subcomplexes of a calculator that are

designed to "hold," "contain," or remember a single number, like
3.1415927 or -19 or 10000. The content of one register is displayed

on the front of your machine: this register is
called the "X-register." However, every calculator has other, invisible registers, and so every
calculator has "memory" in this sense. For
example, let's say you are multiplying 3 by 4.
If you key the 3 into the machine first, 3 will
be the first number visibly displayed. However,
when you then key in 4, 4 will replace 3. Thus 4 becomes the content of the X-register. However, we know that the 3 is still somewhere in the machine, because when multiplication has been performed,
12 is displayed. Thus the machine has shown that it remembered the
3 by multiplying it by the number in X. The invisible register where
3 was held, while 4 was displayed, is called the Y-register. Every
calculator has two registers, X and Y, where the numbers are held
just before the binary operations $\boxed{\times}$ and $\boxed{\div}$ are performed. Often
these same registers are used for $\boxed{+}$ and $\boxed{-}$ also, but on some
machines a third register Z holds numbers z destined to be the addend
in $z + x$ or $z - x$. Still other models have provision for "constant"
multiplication, etc., and then they remember that constant, possibly
in Y or in Z.

The last mentioned machines are examples of those having "algebraic logic." Many scientific machines have "reverse Polish logic";
on these there is a "stack" of three or four registers that are used
in arithmetic: these are called X, Y, Z, and T if there is a fourth
one. Some algebraic machines have parenthesis buttons $\boxed{(}\boxed{)}$, and
possibly $\boxed{[}\boxed{]}$ as well, which correspond to still further invisible
registers. And, of course, any of the above types of machines may
provide a memory register M, or even several memories, in addition
to the registers logically assigned to arithmetic.

The above discussion will be quite confusing to you until you
understand the logic design of your own machine. But there are too
many styles of calculator architecture for this discussion to cover
them all in detail. Use your *Owner's Manual*; study it carefully and

work out the examples there just as they are given. After each keystroke, try to understand which number is in which register (some manufacturers' literature is quite obscure on this point).

Program Records

Here is a system that will aid you in understanding the logical flow of your machine's work. It will also help you to plan a "program" for any given calculation. We illustrate the idea with some simple examples in algebraic logic. The format records the content of each register after keying in DATA or keying a function (="FN").

Calculate 3.4X5.6:

line no.	DATA	FN	PROGRAM RECORD				
			X	Y	Z		Comment
1	3.4		3.4				No entry for the Y
2		X	3.4	3.4			register on a given
3	5.6		5.6	3.4			line means that its
4		=	19.04				content is not relevant.

Calculate (3.4X5.6) + 7.8:

line no.	DATA	FN	PROGRAM RECORD				
			X	Y	Z		Comment
1	3.4		3.4				
2		X	3.4	3.4			
3	5.6		5.6	3.4			
4		+	19.04	19.04			
5	7.8		7.8	19.04			
6		=	26.84				

If your machine has a memory register M into which you may add the content of X by keying $\boxed{\Sigma}$ or $\boxed{M+}$, then you can do two computations simultaneously. One sum can be formed in M while another computation is carried out in the usual way in the arithmetic registers X, Y, and Z. We illustrate this with one example, though the idea may be applied to many different computations.

Calculate $(3.4^2+5.6^2)/(3.4+5.6)$:

line no.	DATA	FN	X	Y	Z	M	Comment
							PROGRAM RECORD
1	3.4		3.4				
2		STO	3.4			3.4	Previous content of M is lost
3		x^2	11.56			3.4	
4		+	11.56		11.56	3.4	Addition uses X & Z in machine with hierarchy
5	5.6		5.6		11.56	3.4	
6		Σ	5.6		11.56	9.	Arith. into M only; no effect on $X,Y,$ or Z
7		x^2	31.36		11.56	9.	
8		=	42.92			9.	Arith. between X & Z; M not disturbed
9		÷	42.92	42.92		9.	
10		RCL	9.	42.92		9.	Content of M is recalled
11		=	4.7688889			9.	

There are two blank "Program Record" sheets included at the end of this volume. You may remove one and photocopy it. This will provide you with blanks on which to study and record programs for your own machine.

REWRITING FORMULAS

Suppose you wish to calculate $(3.4\times5.6)+(7.8\times9.1)$: simple algebraic logic will not handle this problem as it is stated. On machines with a separate memory register M, $3.4\times5.6 = 19.04$ may be computed and stored in M. Then $7.8\times9.1 = 70.98$ is calculated, 19.04 recalled from M and added to 70.98. And, of course, if your machine has no memory register, you may write down the intermediate result 19.04 and later key it in again after finding 7.8X9.1. There is another alternative: rewrite the expression as [(3.4X5.6)÷9.1+7.8]X9.1. This revised expression requires four binary operations, and it also requires that 9.1 be keyed in twice. But it does avoid your having to write down any intermediate results.

We list below some algebraic identities that may be useful in rewriting sums to fit the logical architecture of your machine.

$$(a\text{X}b)+(c\text{X}d) = (a\text{X}b/d+c)\text{X}d \qquad \text{(Sum of Products)}$$

(Note that you may choose the simplest of the four numbers above to be d, the number that must be re-entered.)

$$\frac{a}{b} + \frac{c}{d} = \left(\frac{a\text{X}d}{b} + c\right)/d \qquad \text{(Sum of Quotients)}$$

$$\left(\frac{1}{b} + \frac{1}{d}\right)^{-1} = d/\left(\frac{d}{b} + 1\right) \qquad \text{(Reciprocal of Sum of Reciprocals)}$$

A sum of squares is

$$\sqrt{a^2+b^2} = \sqrt{(a/b)^2 + 1}\ \text{X}b \qquad \text{(Sum of Squares)}$$

Each of these identities may be extended to an expression involving more than two summands.

CONSTANT ARITHMETIC

Many calculators having algebraic logic provide for "constant arithmetic," where repeated use is made of one fixed number in operations on various other numbers. There is a way to realize this mode of operation with some calculators having Polish logic as well. With the desired constant a in the X-register, key $\boxed{\text{ENTER↑}}$ three times to "fill the stack." If another number b is placed in the X-register now and the $\boxed{\text{X}}$ operation performed repeatedly, the numbers

calculated and displayed will be, successively, b, ab, a^2b, a^3b, ...
This is a *geometric progression*: for an *arithmetic progression*, key
$\boxed{+}$ repeatedly instead of $\boxed{X}$, to see b, $a+b$, $2a+b$, $3a+b$, ... Unfortunately, not all machines with Polish logic will endlessly dup-
licate a at the top of the stack; thus this method will not work
with some.

Constant arithmetic is sometimes useful in forming Riemann sums
or partial sums for series. For example, filling the stack with the
number 100 (or using 100 as a constant addend) speeds the calculation of $\sum_{i=0}^{10} \sqrt{100-i^2}$.

Factoring Integers

A calculator will quickly and reliably find the prime factorization
of an integer. Suppose that an integer n is not itself a prime, so
it has factors $n = kl, k \neq 1 \neq l$. Not both k and l can be larger
than $\sqrt{n}$; hence there is some prime factor of n that is less than $\sqrt{n}$.
Accordingly, one may factor any non-prime integer $n \leq 10000$ by testing it for divisibility by all the primes $p < 100$. The list of such
primes is

$$2, \ 3, \ 5, \ 7, \ 11, \ 13, \ 17, \ 19,$$
$$23, \ 29, \ 31, \ 37, \ 41, \ 43, \ 47, \ 53,$$
$$59, \ 61, \ 67, \ 71, \ 73, \ 79, \ 83, \ 89, \ 97.$$

As an example, we factor 13083. Since it is odd, it is not divisible
by 2; we first try $13083 \div 3 = 4361$. Hence 3 is a factor. Next,
$4361 \div 3 = 1453.667$, so 3 is not a repeated factor. Try in succession $4361 \div 5 = 872.2$ and $4361 \div 7 = 623$. Hence 7 is the next factor;
we test $623 \div 7 = 89$, which is a prime. The factorization is complete:
$13038 = 3 \times 7^2 \times 89$.

Integer Parts and Conversion of Decimals

To display the integer part of a number in an 8-digit machine, first
add 10^7 and then subtract 10^7. In a 10-digit machine use 10^9; if
the machine rounds numbers upward at the end of its computations,
first subtract $\frac{1}{2}$, then add 10^9. then subtract 10^9.

To convert decimal yards into yards and feet (and, later, inches), subtract off the integer part of the yards figure and multiply by 3. As an example, 3.456 yards is 3 yards plus 0.456X3 = 1.368 feet. Furthermore, 0.368 feet is 0.368X12 = 4.416 inches, so 3.456 yards = 3 yards, 1 foot, 4.416 inches. For a backwards example, convert 7 yards, 2 feet, 5 inches to 7 yards plus 2 and 5/12 feet = 7 yards, 2.4166667 feet, which is 7 + 2.4166667÷3 = 7.8055556 yards.

The conversions of decimal hours to hours-minutes-seconds and of decimal degrees to degrees-minutes-seconds are handled similarly,

POLYNOMIAL EVALUATION AND SYNTHETIC DIVISION

Let $p(x) = a_0x^n + a_1x^{n-1} + \ldots + a_{n-1}x + a_n$ be a polynomial; the most efficient computation of its value $p(z)$ at a number z is

$$p(z) = \big[[(a_0 \text{X} z + a_1) \text{X} z + a_2] \text{X} \ldots \big] \text{X} z + a_n.$$

A fringe benefit of this method of evaluating polynomials is that it performs a synthetic division simultaneously. At certain stages of the computation the coefficients a_0, $a_0 \text{X} z + a_1$, $(a_0 \text{X} z + a_1) \text{X} z + a_2$, $\ldots$ of a polynomial $q(x) = b_0x^{n-1} + b_1x^{n-2} + \ldots + b_{n-1}$ are displayed, where $q(x)$ is the quotient of $p(x)$ divided by $(x-z)$. That is,

$$p(x) = (x-z)q(x) + p(z).$$

Here $p(z)$, the *evaluation* of $p(x)$ at $x = z$, is the *remainder* after division of $p(x)$ by $(x-z)$. As an example, let

$$p(x) = 2x^3 + 3x^2 + 4x + 5$$

and $z = 6$: first fill the stack and then compute as in Program Record A. This program is written for Polish logic with a four-register stack; easy modifications will adapt it to other machines (with algebraic logic it is easiest to use M to store the number z when it has 8 digits).

$$2x^3 + 3x^2 + 4x + 5 = (x-6)(2x^2+15x+94) + 569$$

The evaluation is as $([(2)6+3]6+4)6+5$

line no.	DATA	FN	X	Y	Z	T	Comment
1	6.		6.				z is entered into X
2		↑	6.	6.			
3	:	↑	6.	6.	6.		
4		↑	6.	6.	6.	6.	stack is filled with z
5	2		2.	6.	6.	6.	$a_0 = b_0$ is entered into X
6		X	12.	6.	6.	6.	content of T is duplicated into Z
7	3		3.	12.	6.	6.	
8		+	15.	6.	6.	6.	b_1 is content of X
9		X	90.	6.	6.	6.	
10	4		4.	90.	6.	6.	
11		+	94.	6.	6.	6.	b_2 is content of X
12		X	564.	6.	6.	6.	
13	5		5	564.	6.	6.	
14		+	569	6.	6.	6.	$p(z)$ is content of X

TAYLOR SERIES EVALUATION

A partial sum of a Taylor series is a polynomial, of course. Thus the method offered above may be most efficient on your machine for evaluating the partial sum $S_5(x) = x - x^3/3! + x^5/5!$ at $x = 0.1234$ in order to approximate

$$\sin 0.1234 \doteq \left[\left(\frac{1}{5!} 0.1234^2 - \frac{1}{3!}\right)X0.1234^2 + 1\right]X0.1234.$$

227

For a machine without a button for $\boxed{n!}$ or for $\boxed{1/x}$, there is a modification of this method that may reduce the total number of arithmetic operations:

$$\sin 0.1234 \doteq \left[\left(0.1234^2/5/4 - 1\right)\text{X}0.1234^2/3/2 + 1\right]\text{X}0.1234.$$

This method may also reduce "round-off" error (see below).

Here is a step-by-step illustration for setting up such expressions:

$$
\begin{aligned}
S_5(x) &= x - x^3/3! + x^5/5! \\
&= x(1 - x^2/3! + x^4/5!) \\
&= x\left(1 + x^2/3!\left[-1 + x^2/5/4\right]\right) \\
&= \left[\left(x^2/5/4 - 1\right)x^2/3/2 + 1\right]x.
\end{aligned}
$$

ARTIFICIAL SCIENTIFIC NOTATION

Some machines will display and calculate with numbers in *scientific notation*; for instance, they display ⌷2345.678 as ⌷.2345678 04 which means $1.2345678\text{X}10^4$. Here the number 1.2345678 will be called the *mantissa*, and 4 is the *exponent*. If your machine does not have scientific notation, there are two circumstances in which you may wish to imitate it by hand. One such case is when you wish to calculate with very large or very small numbers, such as $6.02\text{X}10^{23}$ or $6.4384696\text{X}10^{-7}$, which cannot be entered into the machine at all. Another such circumstance is where your computation results in a very small number. For example, $1.23210 \div 4567891.2$ is given as 2.6973059-07 (which means the number $2.6973059\text{X}10^{-7}$) on a machine with scientific notation, but the answer on other machines is 0.0000002. To avoid such loss of information or to calculate with very large or small numbers, merely use the machine to do the "mantissa arithmetic" while you keep track of the exponents yourself on a sheet of paper. This is easy: add exponents for multiplication

228

and subtract exponents to divide. If an intermediate result with mantissas is a number not near 1, divide or multiply it by an appropriate power of 10 and add or subtract the appropriate integer from your exponent.

Be careful during addition or subtraction of mantissas to rewrite the numbers so that the two rewritten mantissas belong to the same exponent. Some examples are:

$$(6.02 \text{X} 10^{23}) \text{X} (6.4384696 \text{X} 10^{-7}) = 38.759587 \text{X} 10^{16} = 3.8759587 \text{X} 10^{17}$$

$$(6.4384696 \text{X} 10^{-7})^3 = 266.89961 \text{X} 10^{-21} = 2.6689961 \text{X} 10^{-19}$$

$$(1.234 \text{X} 10^{13}) + (5.678 \text{X} 10^{16}) = (0.001234 \text{X} 10^{16}) + (5.678 \text{X} 10^{16})$$

$$= 5.679234 \text{X} 10^{16}$$

ROUND-OFF, OVERFLOW AND UNDERFLOW

In the preceding section we discussed the loss of information in a machine without scientific notation when results are very small numbers. This is one example of *round-off*, which is a type of computational error arising from a machine's inability to represent or display numbers with more than 8 (or 10) *significant digits* (that is, digits of accurate information in the mantissa). This sort of round-off is dealt with effectively by means of artificial scientific notation.

A frequent source of round-off lies in the subtraction of two nearly equal numbers. This occurs inevitably in numerical differentiation, for instance. Various tricks may be used to avoid or mitigate this kind of round-off. One method we have used with differentiation is to evaluate $\lim_{h \to 0} \frac{f(x+h) - f(x-h)}{2h}$ instead of using the usual difference quotient. Another useful technique will help with the problem encountered in Problem P5, Chapter 2, of evaluating

$$\lim_{x \to 0} \frac{67.89^x - 1}{x}.$$

Here we rewrite the fraction, multiplying numerator and denominator by the same number $67.89^x + 1$ to get in succession:

$$\frac{67.89^x - 1}{x} = \frac{67.89^{2x} - 1}{(67.89^x+1)x} = \frac{67.89^{4x} - 1}{(67.89^{2x}+1)(67.89^x+1)x} \ .$$

This modification results in an additional one or two correct digits in the limit, which is ln 67.89. (See Problem P6, Chapter 6 for another derivative approximation.)

Overflow and *underflow* result when computations produce numbers that are too large or too small for the machine to express. An example is 67.89^{75}, which no pocket calculator can handle. One way to work this problem is with artificial scientific notation: $(67.89)^{75} = 6.789^{75}\times10^{75} = (2.4290454\ 62)\times10^{75} = 2.4290454\times10^{137}$. But that will not help when the problem is to calculate 6.789^{175}. For this example, compute

$$6.789^{175} = (6.789^{25})^7 = (6.2394343\ 20)^7 = (6.2394343^7)\times10^{140}$$

$$= (3.6814297\ 05)\times10^{140}$$

$$= 3.6814297\times10^{145}$$

(Here the operations inside the parentheses are carried out by the machine.) Underflow may be handled by similar methods.

HANDLING LARGE EXPONENTS

Here is a more systematic method than those mentioned above for e^x when x is large. Write $e^x = a\times10^b$, where $b = [x/\ln 10]$, the "integer part" of $x/\ln 10$. Consequently, $a = 10^c$ where c is the "fractional part" of $x/\ln 10$. Thus we have $e^x = 10^c\times10^{[x/\ln 10]} = 10^{x/\ln 10}$, which you may check by taking logarithms of e^x and $10^{x/\ln 10}$.

For example, compute e^{1234} by first calculating $1234/\ln 10 = 535.91939$. Hence $[1234/\ln 10] = 535$ and $c = 0.91939$, so $10^c = 8.3059632$. Since the last three digits of 10^c may be in error (do you see why?), we report that $e^{1234} = 8.3059\times10^{535}$.

230

MACHINE DAMAGE AND ERROR

Your calculator has a reliability curve rather like that for a bathtub. That is, like a bathtub, if it is not defective when it is
delivered new, then it is not likely to break down soon. Its buttons,
for example, are designed to last for something like a million clo-
sures. The weak points of most pocket calculators tend to be the
batteries, switches, and display lights, rather than the incredibly
complex, integrated transistor circuitry that does the arithmetic.
However, this circuitry can be damaged or can make errors in arithmetic and memory if it is given an electrostatic shock. This can

occur if you touch the machine just
after walking across a thick rug. It
is also easy to avoid: always ground
yourself first before touching the
calculator. For similar reasons, the
calculator should be OFF when the
adapter is plugged into an outlet or
into the machine.

Some battery chargers are not
adapters, and the batteries can then
be ruined by overcharging; check your
instruction book. In other machines, the batteries will completely
discharge, and may be damaged, if the switch is left ON while the
adapter is connected to the calculator yet not plugged into the
power outlet.

Of course, your calculator will last longer if you do not bang
it or drop it. Also, avoid storing it in such hot places as a car's
glove compartment in summertime.

If you are interested further in the physical and electronic
design of your machine, you will enjoy reading *Electronic Calculators*
by H. Edward Roberts (Indianapolis: Howard W. Sams, 1974).

For a thorough description and pictures of Large Scale Integrated ("LSI") circuits, see the article "Metal-Oxide Semiconductor
Technology" by William Hittinger in *Scientific American*, August, 1973,
pp. 48-57.

REFERENCE DATA AND FORMULAS

GREEK ALPHABET

A	α	Alpha	I	ι	Iota	P	ρ	Rho
B	β	Beta	K	κ	Kappa	Σ	σ	Sigma
Γ	γ	Gamma	Λ	λ	Lambda	T	τ	Tau
Δ	δ	Delta	M	μ	Mu	Y	υ	Upsilon
E	ε	Epsilon	N	ν	Nu	Φ	φ	Phi
Z	ζ	Zeta	Ξ	ξ	Xi	X	χ	Chi
H	η	Eta	O	o	Omicron	Ψ	ψ	Psi
Θ	θ	Theta	Π	π	Pi	Ω	ω	Omega

MATHEMATICAL CONSTANTS

π = 3.1415 92653 58979 32384 62643

π^{-1} = 0.3183 09886 18379 06715 37768

e = 2.7182 81828 45904 52353 60287

e^{-1} = 0.3678 79441 17144 23215 95524

γ = 0.5772 15664 90153 28606 06512

CONVERSION OF UNITS: U.S. - ENGLISH TO S.I. - METRIC

1 inch	= 0.0254^e meter	= 2.54^e centimeters
1 foot	= 0.3048^e meter	= 30.48^e centimeters
1 yard	= 0.9144^e meter	
1 statute mile	= 1609.344^e meters	= 1.609344^e kilo-meters
1 nautical mile	= 1852^e meters	= 1.852^e kilometers
1 acre	= 0.4046856 hectares	= 4046.8564 square meters
1 fluid ounce	= 0.0295735 liters	= 29.574358 cc

^{e}A superscript e indicates that the conversion factor is exact.

```
1 U. S. gallon          = 3.7854118 liters
1 Imp. gallon          = 4.545960 liters
1 ounce (avdp.)        = 0.0283495 kilogram    = 28.349523 grams
1 pound (avdp.)        = 0.4535924 kilogram    = 453.59237ᵉ grams
1 pound (apoth. or troy) = 0.3732417 kilograms = 373.24172 grams
1 pound force          = 4.4482216 newtons
1 slug                 = 14.5939 kilograms
1 poundal              = 0.138255 newtons
1 foot-pound           = 1.35582 joules
1 B.T.U.               = 1055 joules
```

temperature: $(F^\circ - 32)5/9 = C^\circ$

$\qquad\qquad 9C^\circ/5 + 32 = F^\circ$

To convert square or cubic units use the square or cube of the appropriate conversion factor. For example, 1 cubic inch = 2.54^3 cc = 16.387064 cc. To convert metric to English use reciprocal factor.

ALGEBRA

Sum of the Integers: $\quad 1 + 2 + 3 + \ldots + n = \frac{1}{2} n(n+1)$
(Arithmetic Progression)

Sum of the Squares: $\quad 1^2 + 2^2 + 3^2 + \ldots + n^2 = \frac{1}{6} n(n+1)(2n+1)$

Sum of the Cubes: $\quad 1^3 + 2^3 + 3^3 + \ldots + n^3 = \frac{1}{4} n^2 (n+1)^2$

Sum of the Powers: $\quad 1 + r + r^2 + r^3 + \ldots + r^{n-1} = \frac{1-r^n}{1-r}$ if $r \neq 1$
(Geometric Progression)

Difference of Powers: $\quad x^n - y^n = (x-y)(x^{n-1} + x^{n-2}y + x^{n-3}y^2 + \ldots$

$\qquad\qquad\qquad\qquad\quad + xy^{n-2} + y^{n-1})$

Quadratic Formula for
Zeros of $\qquad\qquad ax^2 + bx + c: \quad x = \frac{-b \pm \sqrt{b^2 - 4ac}}{2a}$

Binomial Coefficients: $\quad \binom{n}{r} = \frac{n!}{r!(n-r)!} = \frac{n(n-1)(n-2)\ldots(n-r+1)}{r!}$

Binomial Theorem: $(x+y)^n = \binom{n}{0} x^n + \binom{n}{1} x^{n-1} y + \ldots + \binom{n}{r} x^{n-r} y^r + \ldots$

$$+ \binom{n}{n} y^n$$

GEOMETRY

Triangle Area = $bh/2$

Parallelogram Area = bh

Trapezoid Area = $(a+b)h/2$

Circle Area = πr^2. Circumference = $2\pi r$

Sphere Area = $4\pi r^2$, Volume = $4\pi r^3/3$

Ellipsoid Volume = $4\pi abc/3$

Prism Volume = Bh, where B is base area

Right Circular Cylinder Volume = $Bh = \pi r^2 h$

Pyramid Volume = $Bh/3$

Right Circular Cone Volume = $\pi r^2 h/3$

Point-Slope Line: $y - y_1 = m(x-x_1)$

Slope-Intercept Line: $y = mx + b$

Two-Point Line: $y - y_1 = (x-x_1)(y_2-y_1)/(x_2-x_1)$

General Line: $ax + by + c = 0$

Angle between two lines: $\arctan\left(\dfrac{m_2-m_1}{1+m_2 m_1}\right)$

Parallel lines: $m_1 = m_2$

Perpendicular lines: $m_1 m_2 = -1$

Distance (x,y) to line: $\dfrac{ax+by+c}{(a^2+b^2)^{\frac{1}{2}}}$

Translation of origin to (h,k): $x' = x - h$, $y' = y - k$

Rotation of axes: $x' = x \cos \alpha + y \sin \alpha$, $y' = -x \sin \alpha + y \cos \alpha$

ELLIPSE; CENTER AT ORIGIN

$x^2/a^2 + y^2/b^2 = 1$; $x = a \cos \theta$, $y = b \sin \theta$

Line tangent at (x_1, y_1): $x_1 x/a^2 + y_1 y/b^2 = 1$

If $a > b$: $c^2 = a^2 - b^2$; $e = c/a$; foci $= (\pm c, 0)$

If $b > a$: $c^2 = b^2 - a^2$; $e = c/b$; foci $= (0, \pm c)$

Area: $A = \pi ab$

Hyperbola, Center at Origin

$x^2/a^2 - y^2/b^2 = 1$; $x = a \sec \theta$, $y = b \tan \theta$

Line tangent at (x_1, y_1): $x_1 x/a^2 - y_1 y/b^2 = 1$

Asymptotes: $y = \pm bx/a$

$e = c/a$, foci $= (\pm c, 0)$, where $c = \sqrt{a^2 + b^2}$

Conjugate, center at origin: $y^2/b^2 - x^2/a^2 = 1$; $y \doteq b \sec \theta$,

$$x = a \tan \theta$$

Line tangent at (x_1, y_1): $y_1 y/b^2 - x_1 x/a^2 = 1$

Asymptotes: $y = \pm bx/a$

$e = c/b$, foci $= (0, \pm c)$, where $c = \sqrt{a^2 + b^2}$

Center at (a, b), with asymptotes $x = a$, $y = b$: $(x-a)(y-b) = k$

Parabola, Vertex at Origin, Opening in Direction of Positive y

$y = x^2/4p$, $x = 2pt$, $y = pt^2$

Line tangent at (x_1, y_1): $2p(y+y_1) = x_1 x$

Focus $= (0, p)$. Directrix: $y = -p$

Conic: Eccentricity e, Focus at the Origin, and Corresponding Directrix $x = -k$:

$$r = \frac{ek}{1 - e \cos \theta}$$

Trigonometric Functions

$\sin \pi/6 = \cos \pi/3 = 1/2$ $\qquad$ $\sin \pi/3 = \cos \pi/6 = \sqrt{3}/2$

$\tan \pi/6 = \text{ctn } \pi/3 = \sqrt{3}/3$ $\qquad$ $\tan \pi/3 = \text{ctn } \pi/6 = \sqrt{3}$

$\sin \pi/4 = \cos \pi/4 = \sqrt{2}/2$ $\qquad$ $\tan \pi/4 = \text{ctn } \pi/4 = 1$

$\sin x = 1/\csc x$

$\cos x = 1/\sec x$

$\tan x = 1/\text{ctn } x$

$\sin^2 x + \cos^2 x = 1$

$\tan^2 x + 1 = \sec^2 x$

$1 + \text{ctn}^2 x = \csc^2 x$

$\cos(-x) = \cos x$

$\cos(\pi - x) = -\cos x$

$\cos(\pi + x) = -\cos x$

$\sin(-x) = -\sin x$

$\sin(\pi - x) = \sin x$

$\sin(\pi + x) = -\sin x$

$\sin(x+y) = \sin x \cos y + \cos x \sin y$

$\sin(x-y) = \sin x \cos y - \cos x \sin y$

$\cos(x+y) = \cos x \cos y - \sin x \sin y$

$\cos(x-y) = \cos x \cos y + \sin x \sin y$

$\tan(x+y) = \dfrac{\tan x + \tan y}{1 - \tan x \tan y}$ $\quad$ $(\cos x \cos y \neq 0)$

$\tan(x-y) = \dfrac{\tan x - \tan y}{1 + \tan x \tan y}$ $\quad$ $(\cos x \cos y \neq 0)$

$\sin 2x = 2 \sin x \cos x$

$\cos 2x = \begin{cases} \cos^2 x - \sin^2 x \\ 2 \cos^2 x - 1 \\ 1 - 2 \sin^2 x \end{cases}$

$$\tan 2x = \frac{2 \tan x}{1 - \tan^2 x}$$

$$\left|\sin \frac{x}{2}\right| = \sqrt{\frac{1 - \cos x}{2}} \qquad \left|\cos \frac{x}{2}\right| = \sqrt{\frac{1 + \cos x}{2}}$$

$$\tan \frac{x}{2} = \frac{1 - \cos x}{\sin x} = \frac{\sin x}{1 + \cos x}$$

$$\sin x + \sin y = 2 \sin \frac{x+y}{2} \cos \frac{x-y}{2}$$

$$\sin x - \sin y = 2 \cos \frac{x+y}{2} \sin \frac{x-y}{2}$$

$$\cos x + \cos y = 2 \cos \frac{x+y}{2} \cos \frac{x-y}{2}$$

$$\cos x - \cos y = -2 \sin \frac{x+y}{2} \sin \frac{x-y}{2}$$

$$\sin x \cos y = \frac{1}{2} [\sin(x+y) + \sin(x-y)]$$

$$\cos x \cos y = \frac{1}{2} [\cos(x+y) + \cos(x-y)]$$

$$\sin x \sin y = -\frac{1}{2} [\cos(x+y) - \cos(x-y)]$$

Law of sines: $\dfrac{a}{\sin A} = \dfrac{b}{\sin B} = \dfrac{c}{\sin C}$

Law of cosines: $\quad a^2 = b^2 + c^2 - 2bc \cos A$
$$b^2 = c^2 + a^2 - 2ca \cos B$$
$$c^2 = a^2 + b^2 - 2ab \cos C$$

$$\cos (\sin^{-1} x) = \sin (\cos^{-1} x) = \sqrt{1-x^2}$$

$$\sec (\tan^{-1} x) = \sqrt{x^2+1}$$

$$\sin x = x - \frac{x^3}{3!} + \frac{x^5}{5!} - \ldots + (-1)^{n+1} \frac{x^{2n-1}}{(2n-1)!} + \ldots$$

$$\cos x = 1 - \frac{x^2}{2!} + \frac{x^4}{4!} - \ldots + (-1)^{n-1} \frac{x^{2n-2}}{(2n-2)!} + \ldots$$

EXPONENTIAL AND LOGARITHMIC FUNCTIONS

$a^0 = 1$ $\qquad\qquad\qquad\qquad \log_a xy = \log_a x + \log_a y$

$a^{-n} = 1/a^n$ $\qquad\qquad\qquad a^u = e^{u(\ln a)}$

$a^{1/n} = \sqrt[n]{a}$ $\qquad\qquad\qquad \log_a x \log_b a = \log_b x$

$a^m a^n = a^{m+n}$ $\qquad\qquad\qquad \log_a(a^x) = x$

$(a^m)^n = a^{mn}$ $\qquad\qquad\qquad \log_b a = 1/(\log_a b)$

$(ab)^m = a^m b^m$ $\qquad\qquad\qquad \log_a \frac{r}{s} = \log_a r - \log_a s$

$\log_a x^m - m \log_a x$ $\qquad\qquad \log_b a = \frac{\ln a}{\ln b}$

$$(1 + 1/n)^n < e < (1 + 1/n)^{n+1}$$

$$e^x = \lim_{n \to \infty} (1 + x/n)^n$$

$$e^x = 1 + x + \frac{x^2}{2!} + \frac{x^3}{3!} + \ldots + \frac{x^{n-1}}{(n-1)!} + \ldots$$

$$\ln(1+x) = x - \frac{x^2}{2} + \frac{x^3}{3} - \frac{x^4}{4} + \ldots + (-1)^{n-1} \frac{x^n}{n} + \ldots$$

DIFFERENTIATION

$c' = 0$ $\qquad\qquad\qquad\qquad (cu)' = cu'$

$(u+v)' = u' + v'$ $\qquad\qquad\quad (uv)' = uv' + u'v$

$(u/v)' = \dfrac{vu' - uv'}{v^2}$ $\qquad\qquad (u^n)' = nu^{n-1} u'$

239

Chain Rule: $(u \circ v)' = (u' \circ v)v'$ or $(u[v(x)])' = u'[v(x)]v'(x)$

Inverse Function: $(u^{-1})' = \dfrac{1}{u' \circ u^{-1}}$

$(\cos x)' = -\sin x$ $\qquad\qquad$ $(\sin x)' = \cos x$

$(\cot x)' = -\csc^2 x$ $\qquad\qquad$ $(\tan x)' = \sec^2 x$

$(\csc x)' = -\csc x \cot x$ $\qquad\quad$ $(\sec x)' = \sec x \tan x$

$(\log_a x)' = \dfrac{1}{\ln a}\dfrac{1}{x}$ $\qquad\qquad$ $(\ln |x|)' = \dfrac{1}{x}$

$\qquad\qquad\qquad\qquad\qquad\qquad\quad$ $(e^x)' = e^x$

$(\arcsin x)' = 1/\sqrt{1-x^2}$

$(\arccos x)' = -1/\sqrt{1-x^2}$

$(\arctan x)' = 1/(1+x^2)$

Differential: $dy = f'(x)\ dx$ if $y = f(x)$

INTEGRATION FORMULAS

$$\int cu(x)\ dx = c \int u(x)\ dx$$

$$\int [u(x) + v(x)]\ dx = \int u(x)\ dx + \int v(x)\ dx$$

Integration by Substitution: $\int u[v(x)] \; v'(x) \; dx = U[v(x)]$ if $U' = u$

Integration by Parts: $\int u(x) \; v'(x) \; dx = u(x)v(x) - \int v(x) \; u'(x) \; dx$

Logarithmic Integration: $\int \dfrac{u'(x)}{u(x)} \; dx = \ln \; |u(x)|$

INDEFINITE INTEGRALS (CONSTANTS OF INTEGRATION ARE OMITTED)

$$\int x^n \; dx = \frac{x^{n+1}}{n+1} \text{ if } n \neq -1$$

$$\int \frac{dx}{x} = \ln \; |x|$$

$$\int \frac{dx}{a^2+x^2} = \frac{1}{a} \arctan \frac{x}{a}$$

$$\int \frac{dx}{a^2-x^2} = \frac{1}{2a} \ln \left|\frac{a+x}{a-x}\right|$$

$$\int \frac{dx}{\sqrt{a^2-x^2}} = \text{arc sin } \frac{x}{a}$$

$$\int \frac{dx}{\sqrt{x^2 \pm a^2}} = \ln \; |x+\sqrt{x^2 \pm a^2} \; |$$

$$\int \sqrt{a^2-x^2} \; dx = \frac{1}{2} \left(x\sqrt{a^2-x^2} + a^2 \arcsin \frac{x}{a} \right)$$

$$\int \sqrt{x^2 \pm a^2} \; dx = \frac{1}{2} \left(x\sqrt{x^2 \pm a^2} \pm a^2 \ln \; |x + \sqrt{x^2 \pm a^2}| \right)$$

$$\int x\sqrt{a^2-x^2} \; dx = -\frac{1}{3} (a^2-x^2)^{3/2}$$

$$\int x\sqrt{x^2 \pm a^2} \; dx = -\frac{1}{3} (x^2 \pm a^2)^{3/2}$$

$$\int \frac{xdx}{\sqrt{a^2-x^2}} = -\sqrt{a^2-x^2}$$

$$\int \frac{x\,dx}{\sqrt{x^2 \pm a^2}} = \sqrt{x^2 \pm a^2}$$

$$\int \frac{dx}{x\sqrt{a^2 \pm x^2}} = \frac{1}{a} \ln \left| \frac{a - \sqrt{a^2 \pm x^2}}{x} \right|$$

$$\int \frac{dx}{x\sqrt{x^2 - a^2}} = \frac{1}{a} \arccos \frac{a}{x}, \; x > a > 0$$

$$\int \frac{\sqrt{a^2 \pm x^2}}{x} = dx = \sqrt{a^2 \pm x^2} - a \ln \left| \frac{a + \sqrt{a^2 \pm x^2}}{x} \right|$$

$$\int \frac{\sqrt{x^2 - a^2}}{x}\,dx = \sqrt{x^2 - a^2} - a \arccos \frac{a}{x} \text{ if } 0 < a < x$$

$$\int \sin x \, dx = -\cos x$$

$$\int \sin^2 x \, dx = \frac{1}{2}(x - \sin x \cos x)$$

$$\int \sin^n x \, dx = -\frac{1}{n} \sin^{n-1} x \cos x + \frac{n-1}{n} \int \sin^{n-2} x \, dx$$

$$\int \cos x \, dx = \sin x$$

$$\int \cos^2 x \, dx = \frac{1}{2}(x + \sin x \cos x)$$

$$\int \cos^n x \, dx = \frac{1}{n} \cos^{n-1} x \sin x + \frac{n-1}{n} \int \cos^{n-2} x \, dx$$

$$\int \tan x \, dx = \ln |\sec x|$$

$$\int \tan^2 x \, dx = \tan x - x$$

$$\int \tan^n x \, dx = \frac{1}{n-1} \tan^{n-1} x - \int \tan^{n-2} x \, dx \text{ if } n > 1$$

$$\int \cot x \, dx = \ln |\sin x|$$

$$\int \cot^2 x \, dx = -\cot x - x$$

$$\int \cot^n x \, dx = -\frac{1}{n-1} \cot^{n-1} x - \int \cot^{n-2} x \, dx \text{ if } n > 1$$

$$\int \sec x \, dx = \ln |\sec x + \tan x|$$

$$\int \sec^2 x \; dx = \tan x$$

$$\int \sec^n x \; dx = \frac{1}{n-1} \sec^{n-2} x \, \tan x + \frac{n-2}{n-1} \int \sec^{n-2} x \; dx \text{ if } n > 1$$

$$\int \csc x \; dx = \ln |\csc x - \cot x|$$

$$\int \csc^2 x \; dx = -\cot x$$

$$\int \csc^n x \; dx = -\frac{1}{n-1} \csc^{n-2} x \, \cot x + \frac{n-2}{n-1} \int \csc^{n-2} x \; dx \text{ if } n > 1$$

$$\int \sec x \, \tan x \; dx = \sec x$$

$$\int \csc x \, \cot x \; dx = -\csc x$$

$$\int \sin ax \, \sin bx \; dx = \frac{\sin(a-b)x}{2(a-b)} - \frac{\sin(a+b)x}{2(a+b)} \text{ if } a^2 \neq b^2$$

$$\int \sin ax \, \cos bx \; dx = -\frac{\cos(a-b)x}{2(a-b)} - \frac{\cos(a+b)x}{2(a+b)} \text{ if } a^2 \neq b^2$$

$$\int \cos ax \, \cos bx \; dx = \frac{\sin(a-b)x}{2(a-b)} + \frac{\sin(a+b)x}{2(a+b)} \text{ if } a^2 \neq b^2$$

$$\int \frac{dx}{\sin x \, \cos x} = \ln |\tan x|$$

$$\int \arcsin x \; dx = x \arcsin x + \sqrt{1+x^2}$$

$$\int \arccos x \; dx = x \arccos x - \sqrt{1-x^2}$$

$$\int \arctan x \; dx = x \arctan x - \ln\sqrt{1+x^2}$$

$$\int e^{ax} dx = \frac{1}{a} e^{ax}$$

$$\int a^x dx = \frac{a^x}{\ln a} \text{ if } a > 0, \; a \neq 1$$

$$\int \ln x \; dx = x \ln x - x$$

$$\int (\ln x)^n dx = x(\ln x)^n - n \int (\ln x)^{n-1} dx$$

$$\int \log_a x \; dx = \frac{1}{\ln a} (x \ln x - x)$$

$$\int x^n \ln x \ dx = x^{n+1} \left[\frac{\ln x}{n+1} - \frac{1}{(n+1)^2} \right], \ n \neq -1$$

$$\int x^n (\ln x)^m dx = \frac{x^{n-1}}{n+1} (\ln x)^m - \frac{m}{n+1} \int x^n (\ln x)^{m-1} dx$$

$$\int x e^{ax} dx = \frac{(ax-1)e^{ax}}{a^2}$$

$$\int x^n e^{ax} dx = \frac{x^n e^{ax}}{a} - \frac{n}{a} \int x^{n-1} e^{ax} dx$$

$$\int \frac{e^{ax}}{x^n} dx = - \frac{e^{ax}}{(n-1)x^{n-1}} + \frac{a}{n-1} \int \frac{e^{ax}}{x^{n-1}} dx \ \text{if} \ n > 1$$

$$\int e^{ax} \sin bx \ dx = \frac{e^{ax}(a \sin bx - b \cos bx)}{a^2+b^2}$$

$$\int e^{ax} \cos bx \ dx = \frac{e^{ax}(a \cos bx + b \sin bx)}{a^2+b^2}$$

BIBLIOGRAPHY

Elementary Calculator Manipulation

There are available several introductory guides that teach the use of
calculators in grocery shopping and other elementary applications.
One of these books also augments owners' manuals. It is

> *Slide Rule, Electronic Hand-Held Calculators, and Metrification
> in Problem Solving* by Beakley and Leach (New York: Macmillan,
> 1975).

Two elementary books, which motivate computational skills and numerical understanding by means of elementary arithmetic and number-
theoretic patterns, are

> *Puzzles for a Hand Calculator* by Wallace Judd (Menlo
> Park, CA: Dymax, 1974)

> *The Calculating Book* by James Rogers (New York: Random House,
> 1975).

Advanced Calculator Manipulation

There exists a compilation of numerical techniques adapted for calculators, complete with many detailed button sequences and discussions of easy ways to calculate things. This book offers no theoretical understanding, and its treatments of errors are inadequate.
Nevertheless, those who use their machines heavily in scientific
computation may profit from

> *Scientific Analysis on the Pocket Calculator* by Jon Smith (New
> York: John Wiley and Sons, 1975).

One manufacturer of calculators publishes handbooks that definitely
extend and supplement its owner's manuals. Each of these books is

designed to be used with a specific machine model. Nevertheless, there is much of general interest in one of these:

HP-45 Applications Book (Hewlett-Packard, 1974).

This book contains some repetitious trivia and no theoretical discussions. Yet it also gives recipes for calculations involving complex numbers and complex functions of a complex variable, linear algebra, curve fitting and statistics, number theory, financial calculations, and numerical methods. These recipes frequently exploit the existence of multiple memories in the HP-45, but they may be transcribed to fit other models.

NUMERICAL CALCULUS

Most calculus texts contain discussions of the theory underlying the numerical methods we have used. However, we cite three such texts for their particularly thorough and lucid treatments of real numbers and of numerical integration:

Introduction to Calculus and Analysis, vol. 1, by Courant and John (New York: Interscience, 1965)

University Mathematics by Robert C. James (Belmont, CA: Wadsworth, 1963)

Calculus by Lynn Loomis (Reading, MA: Addison-Wesley, 1974).

A beautiful treatment of both theoretical and numerical aspects of series is given in

Theory and Application of Infinite Series by Konrad Knopp (New York: Hafner, 1947).

There are treatments of the calculus that introduce the use of computers in numerical examples. Their discussions are parallel to ours, although there are surprising differences of emphasis.   Two such books are:

> *Computer Applications for Calculus* by Dorn, Hector, and Bitter (Boston: Prindle, Weber & Schmidt, 1972)

> *Calculus with Computer Applications* by Lynch, Ostberg, and Kuller (Lexington, MA: Xerox, 1973).

NUMERICAL ANALYSIS

George Forsythe has written a very readable essay introducing the numerical aspects of "Solving a Quadratic Equation on a Computer." You can find this essay in

> *The Mathematical Sciences*, COSRIMS (Cambridge: MIT Press, 1969), pp. 138-152.

Advanced students may find the answers to many of their questions in textbooks of "numerical analysis." Some suitable references are:

> *Introduction to Numerical Analysis*, 2nd ed., by F. B. Hildebrand (New York: McGraw-Hill, 1974)

> *Numerical Methods* by Robert Hornbeck (New York: Quantum, 1975)

> *A Survey of Numerical Mathematics*, vol. 1, by Young and Gregory (Reading, MA: Addison-Wesley, 1972).

Some references for more specialized topics are:

> *Methods in Numerical Integration* by Davis and Rabinowitz (New York: Academic Press, 1975)

> *Computer Evaluation of Mathematical Functions* by C. T. Fike (Englewood Cliffs, N.J.: Prentice-Hall, 1968)

Iterative Methods for the Solution of Equations by Joseph Traub
(Englewood Cliffs, N.J.: Prentice-Hall, 1964).

HANDBOOKS

There are two popular handbooks of tables. Each contains much assort-
ed reference information in addition to tabulations of values for many
functions. The first of these is broader and at a lower level. The
second contains formulas, graphs, and tables of values for many functions, plus a great deal of condensed information and guidance on
numerical analysis.

Standard Mathematical Tables, 22nd ed., edited by Samuel Selby
(Cleveland: CRC Press, 1974)

Handbook of Mathematical Functions, edited by Abramowitz and
Stegun (Washington, D.C.: National Bureau of Standards, 1972).
Also published in paperback (New York: Dover, 1972)

APPLICATIONS TO OTHER FIELDS

Most calculus texts discuss applications of this theory outside of
mathematics. Here are several books that devote more than usual attention to these applications:

Mathematics for Life Scientists by Edward Batschelet (New York:
Springer, 1973)

Elementary Quantitative Biology by C. S. Hammen (New York: John
Wiley and Sons: 1972)

Mathematical Methods for Social and Management Scientists by
T. Marll McDonald (Boston: Houghton Mifflin, 1974)

Calculus and Analytic Geometry by Sherman Stein (New York:
McGraw-Hill, 1973)

A Primer of Population Biology by Wilson and Bossert (Stamford: in Conn. Sinaur Associates, 1971).

Journal Suggestions for Students

General	*Nature*
	Science
	Scientific American
Biology	*Journal of Experimental Biology*
Chemistry	*Journal of Chemical Education*
Economics	*Econometrica*
Mathematics	*American Mathematical Monthly*
	Mathematics Magazine
Physics	*American Journal of Physics*
	Physics Today
	American Physicist
Psychology	*Journal of Mathematical Psychology*
	(*Handbook of Mathematical Psychology*, vol. 1,
	2, 3 and 4)

Further Readings in Mathematics

Selected Papers on Calculus edited by Apostal et al. (Belmont, CA: Dickenson, 1968)

The History of the Calculus and its Conceptual Development by Carl Boyer (New York: Dover, 1959)

Number, The Language of Science by Tobias Dantzig (Garden City, NY: Doubleday Anchor, 1954)

Mathematical Thought from Ancient to Modern Times by Morris Kline (New York: Oxford, 1972)

The World of Mathematics by James R. Newman (New York: Simon and Schuster, 1956)

INDEX

line no.	DATA	FN	PROGRAM RECORD				
			X	Y	Z		Comment

line no.	DATA	FN		PROGRAM RECORD			
			X	Y	Z		Comment